Springer Series on Cultural Computing

Editor-in-Chief

Ernest Edmonds, Institute for Creative Technologies, De Montfort University, Leicester, UK

Editorial Board

Cultural Computing is an exciting, emerging field of Human Computer Interaction, which covers the cultural impact of computing and the technological influences and requirements for the support of cultural innovation. Using support technologies such as location-based systems, augmented reality, cloud computing and ambient interaction researchers can explore the differences across a variety of cultures and provide the knowledge and skills necessary to overcome cultural issues and expand human creativity.

This series presents the current research and knowledge of a broad range of topics including creativity support systems, digital communities, the interactive arts, cultural heritage, digital museums and intercultural collaboration.

More information about this series at http://www.springer.com/series/10481

Michiko Ohkura

Editor

Kawaii Engineering

Measurements, Evaluations, and Applications
of Attractiveness

 Springer

Editor
Michiko Ohkura
Shibaura Institute of Technology
Tokyo, Japan

ISSN 2195-9056 ISSN 2195-9064 (electronic)
Springer Series on Cultural Computing
ISBN 978-981-13-7966-6 ISBN 978-981-13-7964-2 (eBook)
https://doi.org/10.1007/978-981-13-7964-2

This Springer imprint is published by the registered company Springer Nature Singapore Pte Ltd.
The registered company address is: 152 Beach Road, #21-01/04 Gateway East, Singapore 189721, Singapore

Preface

"Kawaii" is a Japanese word that denotes "cute," "lovable," or "charming." Although I myself have loved kawaii things ever since I was a child, I had never thought about research kawaii. However, when my son rejected the new alarm clock I had bought for him because it was not kawaii, I was suddenly confronted with a paradigm shift: kawaii is more important than function. This alarm clock experience fueled my kawaii engineering research.

I received B.S. and M.S. degrees as well as a Ph.D. in engineering from the University of Tokyo, and I am currently a professor in the College of Engineering at the Shibaura Institute of Technology. Because research subjects in the field of engineering are artifacts, such as industrial products and services, I have been researching the visibilities, operability, safety, comfort, and the excitement (wakuwaku in Japanese) of them as a researcher of their interfaces. However, because of the reason described above, kawaii became one of my main research interests.

My Ph.D. advisor was Prof. Susumu Tachi, a leading Virtual Reality (VR) researcher, and the title of my 1995 thesis was "Research on the auditory spatial perception characteristics of human beings." In it, I used VR as a tool for elucidating human perception characteristics, a practice I have maintained in the subsequent decades on both human perception characteristics and emotion. Many of the kawaii engineering studies introduced in this book also utilize VR and Augmented Reality (AR). Many kawaii interactive works are the research results of young researchers who are active in the Japan Virtual Reality Society. Although they are not subject to kawaii research, they plan to "make it kawaii" when they shape their research results, which are popular with non-Japanese researchers at international conferences and exhibitions. This situation suggests the universality of the Kansei/affective values of kawaii.

For this book, which focuses on the kawaii engineering research, I have been doing for more than 10 years, and I have asked some leading researchers in the field to submit chapters.

In Chap. 1, I discuss the background of my initial research on kawaii engineering in 2006 and outline this book.

Chapter 2 introduces experiments on the systematic measurement and evaluation methods for kawaii products and individually address the physical attributes of color, shape, material perception (visual texture, tactile texture), and sound. The experimental participants were mainly males and females in their 20s, and such relatively young people participated without embarrassment. We, students and I, primarily used a virtual environment to present objects. Real products are composed of many combinations of physical attributes, and their interaction determines the kawaii impression. However, as engineering research, we conducted a series of such studies, because it is critical to conduct fundamental experiments and isolate the "kawaii feeling" caused by a single physical attribute in the early stage of research. Even though individual conclusions may not be so surprising, I believe that more important than the conclusions themselves is the process that can clarify the relation between the kawaii feeling and each attribute by taking a systematic engineering approach. In addition, although we employed virtual objects for our experiments, experiments with real objects have become possible because of the recent availability of 3D printers.

Chapter 3 introduces affective evaluation experiments on both the visual and tactile material perceptions of bead-coated resin surfaces that systematically expanded the materials, the diameters, and the hardness of the beads.

In Chap. 4, Dr. Sugano and Dr. Tomiyama propose a kansei mathematical model and describe their attempts to identify the physical attributes that determine kawaiiness in motion.

Chapter 5 introduces research on kawaii feelings using biological signals. After introducing the advantages of measuring biological signals over such subjective evaluations as questionnaires, we explain the position of positive and dynamic feelings in Russell's circumplex model and introduce our research results of wakuwaku and kawaii feelings as well as the classification of kawaii stimuli into exciting and relaxing aspects.

Chapter 6 introduces our eye-tracking research that identifies the relationship between kawaii feelings derived from kawaii illustrations and eye movements.

In Chap. 7, Dr. Nittono briefly describes recent trends in psychological research on kawaii and cuteness.

Chapter 8 introduces the attempt to increase the appetite of senior citizens by kawaii spoons. Their mental states were estimated from ECG readings.

Chapter 9 introduces the attempt for the practical implementation of an emotion-driven digital camera by EEG.

In Chap. 10, we evaluate the key emotional values that influence Saudi women to purchase luxury fashion brands. In addition, we define kawaii from their perspective and identify the key desired emotional values for their importance compared to kawaii values. For both studies, we employed a qualitative method to gather a comprehensive understanding of the subjects.

Chapter 11 introduces a mathematical model of kawaii spoon evaluations by Thai and Japanese participants using Support Vector Machine (SVM).

In Chap. 12, we use a deep Convolutional Neural Network (CNN) for a mathematical model of kawaii cosmetic bottle evaluations by Thai and Japanese participants. Our results clarified the effective attributes for increasing the kawaiiness and the effectiveness of our constructed model.

Chapter 13 introduces our experimental research on the kawaii perception of artifacts by Chinese and Japanese people. Our results indicate that culture, gender, and age affect kawaii perceptions.

In Chap. 14, Shinsuke Ito introduces how a group of people with different backgrounds teamed up to create a very cute car named "RimOnO" and how their prototype matched its current design.

Chapter 15 summarizes this book and the future of kawaii engineering.

In 2017, my Japanese book entitled *Kawaii Kogaku* (Kawaii Engineering) was published. I wrote and edited it in 2016, which was deemed by some to be a banner year for virtual reality. Because much of my research on kawaii engineering was performed using virtual environment, I experienced conjunction between its publication and VR's rapid spread. In 2018, the Japan Society of Kansei Engineering (JSKE), of which I am a vice president, held its 20th memorial annual meeting at The University of Tokyo. I led some events related to kawaii engineering at the meeting. I hope this book fuels both the continued advancement of kawaii engineering and JSKE.

I am confident that readers will enjoy the new optimistic research field called kawaii engineering. I thank my co-researchers as well as the students who participated in the research. I also thank the writers and researchers who contributed to this book and the Springer staff who edited it.

Tokyo, Japan   Michiko Ohkura

Contents

Part I
Introduction

Chapter 1
Kawaii Engineering

Michiko Ohkura

Abstract This chapter introduces the background of the birth of kawaii engineering in Japan and describes this idea's objectives. The second section of this chapter introduces kawaii's cultural background and describes the cultural differences between Europe/US and Japan. Finally, the last section introduces our preliminary experiment in 2007 into kawaii engineering research.

Keywords Kawaii engineering · Kansei value · Affective value · Kawaii · Fancy goods · *Pillow book* · Shape · Sympathy · Peace · SDGs

1.1 What Is Kawaii Engineering?

This chapter is mainly based on Refs. [1–3].

The rapid progress of science and technology in the twentieth century ushered in a materially affluent society. The striking development of information and communication technologies in its latter half provided such incredibly convenient tools as computers and network environments. Against this backdrop, people in the twenty-first century tend to place more importance on spiritual wealth than material wealth and have modified their value systems from physical-based to information-based ones (e.g., [4]). Moreover, to break through the stagnation that threatened Japanese manufacturing, the Japanese government selected Kansei (emotion or affection) to be the fourth product value, joining the values of function, reliability, and cost [5].

The following is our research question: "What form would human-friendly information take?" We research under the rubric that such socially vulnerable people as women, children, the physically challenged, and the elderly must be supported by the technologies of the twenty-first century's advanced information societies. For example, our research clarified the conditions of living/working spaces using Kansei experiments for securing safe, comfortable, and high-quality life especially for the elderly and women (e.g., [6–8]). In parallel, we developed a control system using

M. Ohkura (✉)
Shibaura Institute of Technology, Tokyo, Japan
e-mail: ohkura@sic.shibaura-it.ac.jp

© Springer Nature Singapore Pte Ltd. 2019
M. Ohkura (ed.), *Kawaii Engineering*, Springer Series on Cultural Computing, https://doi.org/10.1007/978-981-13-7964-2_1

3

a mechanical pet named AIBO, manufactured by Sony from 1999 to 2006, which moves by judging the user's alpha waves in a way that makes the user feel the most comfortable [9–11]. From the results of these researches, the Kansei values people feel from spaces and mechanical pets, including a sense of safety and comfort, can be quantitatively identified by questionnaires and the analytical results of biological signals. Then we continued the quantification of Kansei values.

On the other hand, in the advanced information society of the twenty-first century and its communication infrastructure of computers and networks, enriching software that utilizes new technology is critical: that is, digital contents (e.g., [12]). Not only the communication infrastructure but also various amounts of useful knowledge accumulated by highly developed sciences and technologies at present must be increasingly utilized in manufacturing and other fields.

We began our new research endeavor to apply our previous results to the systematic creation of the Kansei values of industrial products. The large export surpluses of Japanese games, cartoons, and animations, so-called digital content [13], exist in contrast with the tremendous import surpluses of other software products [14, 15]. For example, Hayao Miyazaki's animated films such as *Witch's Delivery Service*, *Totoro*, and *Spirited Away* are highly esteemed worldwide and have attracted audiences of millions of people. Such TV animations as Dragon Ball and Doraemon are also being aired in various countries, and Pikachu and other Pocket Monsters are enthusiastically embraced by millions of children all over the world. Furthermore, in July 2016, Pokemon Go was unleashed worldwide and became a social phenomenon. One main reason for the success of those Japanese-born digital contents is the existence of kawaii characters and their highly sensitive techniques [16]. Even though kawaii is a Japanese word that denotes cute, lovable, or charming, we directly use it for the following reasons:

- Kawaii is not exactly the same as cute and/or lovable [17].
- Its use has gradually become internationally recognized and accepted [17, 18].
- It is explained as "(in the context of Japanese popular culture) cute" in the Oxford Dictionary [19].

The differences between kawaii and cute will be described later.

For over 20 years, various Japanese characters, such as Hello Kitty and Pokemon, have spread worldwide. The word kawaii has become part of the international lexicon. Pink digital cameras and round printers are sometimes marketed as kawaii products [20, 21]. However, these productions were ad hoc applications of a particular designer's implicit knowledge, questionnaire results from female high school students, or the taste of charismatic fashion models; they were not the results of a systematic approach focused on kawaii products. Until in 2006, no research has focused on the kawaii attributes of industrial products themselves.

Therefore, we began to systematically analyze the kawaii evaluation of industrial products themselves, because kawaii is created by such attributes as shape, color, visual texture, and tactile texture. We aim to clarify a method for constructing kawaii products based on our results. This is the birth of kawaii engineering in 2007.

1.2 Cultural Background of Kawaii

After being unable to find previous research works related to kawaii in science and technology, we widened our survey's scope and found the following studies on kawaii in the cultural theory field:

- M. Shimamura: Research of fancy—"kawaii" dominates people, goods and money—(1991) [22]
  Shimamura, a Japanese female writer, introduces the social role underlying fancy goods that surround personal belongings and explains how they are required for particular periods. Her foresight and relevance are noteworthy.
- S. Kinsella: Cuties in Japan, Women, Media and Consumption in Japan (L. Skov and B. Moeran, ed.) (1995) [23]
  In this useful reference to Japanese culture from a non-Japanese female perspective, Kinsella broaches such subjects as handwritten letters, spoken words, fancy goods, fashion, food, and pop idols. However, the relationship between these objects and the value of kawaii is not directly argued.
- K. Belson, and B. Bremner: Hello Kitty: The Remarkable Story of Sanrio and the Billion Dollar Feline Phenomenon (2004) [16]
  This book describes Japanese business theory from the viewpoint of American male journalists and focuses its discussion on Hello Kitty and related kawaii characters.
- I. Yomota: "Kawaii" Ron (Kawaii Theory) (2006) [24]
  This book written by a Japanese male researcher on comparative literature addresses the word "kawaii," but basically dismisses it as an adjective used by/for women.

These works recognize the following common attributes of kawaii:

- It is a Kansei/affective value of Japanese origin.
- It denotes such positive meaning as cute, lovable, and charming.

Although these features may seem obvious, such recognition is very misleading because kawaii is also related to the Japanese word "kawaisou," which is generally translated as pitiful in English [25]. In addition, not only pitiful, but also immature, weakness, and incomplete might have negative connotations in Western (Europe and the USA) cultures. However, these words are essentials to the idea of kawaii because matured and complete are valuable in Western cultures; Japanese culture prefers immaturity, such as edible flower buds and cherry blossom buds. The Japanese generally appreciate immature objects because they evoke such feelings as "I want to protect it." That is, immature, weakness, and incomplete are not completely negative, but they are positive Kansei/affective values from a Japanese cultural perspective [26]. According to Endo [26], a Japanese female sociologist, cute is a positive word that is derived from acute, but kawaii includes negative aspects from the viewpoint of Western culture. She describes that this is one large difference between cute and kawaii. Most people in the world have some weak points, and sympathy to kawaii suggests that they can be sympathized with. Therefore, the Japan-originated word kawaii and its concept might be shared all over the world and will encourage peace and sustainable development goals (SDGs) [27] in the twenty-first century.

In addition, A. D. Cheok et al. argued that Japanese kawaii embodies a special kind of cute design, which reduces fear and makes dreary information more acceptable and appealing [28].

Another detailed examination about cute and kawaii was written by Kurosu [29].

According to Yomota [25], the description of the values of kawaii first appeared in Section 151 of the *Pillow Book* written by Sei Shonagon, a famous Japanese female writer, around the year 1000 [30]. She lists some examples of kawaii objects, including the behavior of a chipping sparrow, a small hollyhock leaf, and a cerulean jar. Another article [24] introduced some examples of kawaii industrial products from the Jomon Period, which is more than 2300 years ago. Therefore, kawaii has been a Japanese Kansei value over 2000 years.

Kawaii widely nourished Japanese culture for more than 2000 years from 400 B.C to the Edo Period (1603–1868) [31]. In addition to the kawaii objects described in the *Pillow Book*, we can also find many examples in Ukiyoe and Netsuke in the Edo Period. The "Kawaii Edo Painting" exhibition in Tokyo in the spring of 2013 [32] also suggests that kawaii established as a Japanese Kansei value in the Edo Period.

However, since the Meiji Period (1868–1912), many targets of the adjective kawaii have tended to be limited to the following:

- children, girls, small animals, and other living creatures;
- the facial expressions and gestures of the above creatures;
- imitations of the above creatures (e.g., figures, stuffed animals);
- the facial expressions and/or motions of the above imitations.

These usages of kawaii resemble cute in English and detract from the reality of the awareness that kawaii represented the value of industrial products for more than 100 years from the beginning of the Meiji Period. This phenomenon is strongly connected to Japan's attempts to economically and culturally catch up with Western countries following the Meiji Period. Japanese tend to place more importance on Western values than original Japanese values. This cultural trend reduces awareness to kawaii.

However, based on a cultural background with a long history in Japan and its above concept, kawaii is a unique cultural property of Japan and will spread worldwide in the near future to people in the world who desire peace and SDGs.

1.3 Preliminary Experiment for Kawaii Engineering

1.3.1 Overview

In the book written by I. Yomota, the word kawaii was treated as an adjective only for women (the same as cute) and some descriptions reflect a male-centered viewpoint [25]. For example, "Many girls buy kawaii goods for kawaii appearance, and they are waiting for the moment they are called kawaii by boys." (p. 156). So, we decided to perform preliminary experiments to clarify whether the usage of kawaii in modern Japan suffers from such limitations [25] or we can use kawaii for industrial products as described in the *Pillow Book* [30] and Shimamura's work [23].

The following is our experiment's hypothesis:

Japanese men, especially middle-aged and older, tend to disregard kawaii as an adjective only for living creatures and such figures and characters. On the other hand, Japanese women bestow feelings of kawaii not only on living creatures but also on industrial products.

We tested this hypothesis in 2007 by performing the following experiment [1].

1.3.2 Method

We showed our participants four magnets made of the same materials (metal and rubber) but in different shapes (Fig. 1.1) and gave them the following questionnaire:

1. Put these four magnets in order from the viewpoint of kawaii. If you cannot determine the order for some or all of the magnets, explain why.
2. Evaluate their kawaii degree on a 10-point scale. If you cannot evaluate some or all of them, explain why.
3. Write the scores and your reasons for them. If you do not have a particular reason, no comment is fine.

1.3.3 Results

Twenty participants (10 males and 10 females) in their early 20s and twenty participants (10 males and 10 females) in their early 50s served as volunteers. The numbers of participants who could not place in order the magnets or evaluate some of them are shown in Table 1.1. Assuming these evaluations received 0 points, we calculated the average scores for each magnet (Fig. 1.2).

We reached the following conclusions based on these results and the explanations on the answer sheets:

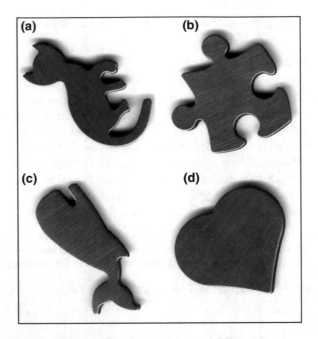

Fig. 1.1 Four kinds of magnets with identical materials and different shapes

Table 1.1 Numbers of participants who could not place in order or evaluate some magnets

	Males in their 50s	Females in their 50s	Males in their 20s	Females in their 20s
A	0	0	0	0
B	3	2	2	2
C	0	0	0	1
D	2	0	2	0

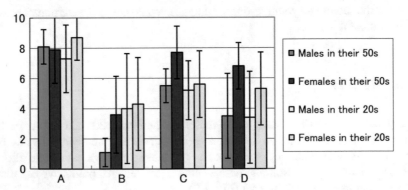

Fig. 1.2 Average magnet scores

1. Magnet A (cat-shaped) got high kawaii scores from all the categories of participants. The following reasons for the high scores were shared among the young and the old and genders:

   - Because I like cats.
   - Cats are kawaii.

2. Magnet B (jigsaw-puzzle-piece-shaped) got the lowest kawaii scores of the four, especially from the male participants in their early 50s. As for the males and females in their early 20s, the evaluations fell into high and low groups. Two reasons were given for the high scores:

   - It is good because it is normal.
   - It is rather exquisite.
     Just one reason was given for the low scores:
   - The puzzle piece is not kawaii.

3. Magnet C (whale-shaped) got the second highest scores on average, especially from females in their early 50s. They gave the following two reasons:

   - Because it is an animal.
   - Because it is a whale.

4. Magnet D (heart-shaped) got the most varied scores between males and females. The scores of the males were very low. The reason was the same for both the high and low scores:

   - Because it is a heart.

A male in his early 50s complained that he could not tolerate to look a heart shape at his age.

1.3.4   Discussion

Male participants in their early 50s gave high scores to the cat- and whale-shaped magnets, suggesting they only felt that living creatures are kawaii. As for males in their early 20s, although their kawaii scores for the heart-shaped magnet are low, their kawaii scores for the puzzle-piece-shaped magnet resemble those of the females, suggesting that males in their early 20s are less affected by the shape of living creatures than males in their early 50s. On the other hand, females gave high kawaii scores to the cat-shaped, the whale-shaped, and the heart-shaped magnets. Females in their early 50s gave relatively high kawaii scores to all the magnets. As for females in their early 20s, the kawaii scores of the whale- and heart-shaped magnets are similar, and some male and female participants in their early 20s and females in their early 50s gave high scores to the puzzle-piece-shaped magnet. From these results, our participants (except males in their early 50s) felt that both animal-shaped and artificial-object-shaped magnets are kawaii. We believe these results verify our hypothesis.

Younger Japanese people embrace kawaii products, but middle-aged Japanese males seem to resist them. Even if kawaii products do not become popular in the short term, perhaps they will spread in the future.

1.4   Conclusion

In 2007, we began our kawaii engineering research, based on more precise consideration of the target attributes of industrial products. In this study, we focused on kawaii as a Kansei value for future industrial products and systematically analyzed its attributes for constructing kawaii products. This step is the beginning of kawaii engineering.

We surveyed cultural studies and confirmed the long history of kawaii as a Kansei value in Japan.

In addition, to support the future possibility of kawaii as a Kansei value, we formed the following hypothesis:

> Japanese men, especially middle-aged and older, tend to treat kawaii as an adjective only for living creatures and such figures and characters. On the other hand, Japanese women feel that kawaii is not only for living creatures but also for artificial objects.

We tested this hypothesis by performing with participants of different ages and genders a simple experiment with kawaii using four different-shaped magnets. The kawaii scores for each magnet differed by age and gender, and we verified the above hypothesis.

This observation suggests that even if the short-term prospects of kawaii products might be disappointing, a strong possibility exists that they will spread in the future.

This book introduces our research results from more than the last 10 years and adds work by other researchers.

Acknowledgements This research was partly supported by the SIT Research Promotion Funds. We thank the students of Shibaura Institute of Technology who contributed to the research and served as volunteers.

References

1. Ohkura, M., & Aoto, T. (2007). Systematic study for "Kawaii" products. In *Proceedings of the 1st International Conference on Kansei Engineering and Emotion Research 2007 (KEER 2007)*, Sapporo.
2. Ohkura, M. (2017). Chapter 1, What is Kawaii engineering. In Ohkura, M. (Ed.), *Kawaii Kogaku (Kawaii Engineering)* (pp. 1–4). Tokyo: Asakura Shoten. (in Japanese).
3. Ohkura, M. (2017). Chapter 2, Cultural background. In Ohkura, M. (Ed.), *Kawaii Kogaku (Kawaii Engineering)* (pp. 5–8). Tokyo: Asakura Shoten. (in Japanese).

4. Ohkura, M. (2005). Interface for human-machine interaction. *Trends in the Sciences, 10*(8), 765–781. (in Japanese).
5. Ministry of Economy, Trade and Industry. Kansei and Value Initiative, Today's New Topics. Retrieved May 22, 2007, from http://www.meti.go.jp/english/newtopics/data/n070522e.html.
6. Ohkura, M., et al. (2005). Comparison of the impression of the space between virtual environment and real environment. In *Proceedings of the 11th International Conference on Human-Computer Interaction 2005.*
7. Aoto, T., et al. (2005). Assessment of relief level of living space using immersive space simulator. *Technical Report of IEICE 2006, 105*(684), 31–34. (in Japanese).
8. Ohkura, M., et al. (2007). Evaluation of comfortable spaces for women using a virtual environment. In *Proceedings of the 12th International Conference on Human-Computer Interaction 2007.*
9. Ohkura, M., et al. (2005). Brain-wave-based motion control system for a mechanical pet. In *Proceedings of the 11th International Conference on Human-Computer Interaction 2005.*
10. Ohkura, M., & Oishi, M. (2006). An alpha-wave-based motion control system for a mechanical pet. *Kansei Engineering International, 6*(2), 29–34.
11. Ohkura, M., & Takano, Y. (2007). An alpha-wave-based motion control system of a mechanical pet for mental commitment. In *Proceedings of 2007 IEEE/ICME International Conference on Complex Medical Engineering-CME 2007.*
12. Ministry of Posts and Telecommunications. Shape of Advanced Information Telecommunications Society and the Role of Government in the 21st Century. Retrieved May 31, 1999 from http://www.soumu.go.jp/joho_tsusin/policyreports/japanese/telecouncil/yakuwari/v21–9905.html. (in Japanese).
13. Japan Electronics and Information Technology Industries Association. Statistics of Exports/Imports of Software in 2000. Retrieved July 31, 2002 from http://it.jeita.or.jp/statistics/software/2000/index.html. (in Japanese).
14. Japan Information Technology Services Industry Association. Press Release. Retrieved October 19, 2005 from http://www.jisa.or.jp/pressrelease/2005-1019.html. (in Japanese).
15. Commerce and Information Policy Bureau, Ministry of Economy, Trade and Industry. (2005). *Digital Content White paper 2005.* Tokyo: Digital Content Association of Japan.
16. Belson, K., & Bremner, B. (2004). *Hello Kitty: The remarkable story of Sanrio and the billion dollar feline phenomenon.* New Jersey: Wiley.
17. Asahi Shinbun. Kawaii (in the morning edition, January 1, 2006). (in Japanese).
18. Wikipedia. Cuteness in Japanese Culture http://en.wikipedia.org/wiki/Cuteness_in_Japanese_culture.
19. Oxford Dictionaries, https://en.oxforddictionaries.com/definition/kawaii.
20. Fuji film Corporation. FinePix Z3 http://www.fujifilm.com/products/digital_cameras/z/finepix_z3/index.html.
21. Seiko Epson Corporation. Colorio me http://www.epson.jp/osirase/2004/040316_1.htm. (in Japanese).
22. Shimamura M. (1991). *A research on fancy—"kawaii" dominates humans, things, and money-.* Tokyo: Nesuko. (in Japanese).
23. Kinsella, S. (1995). Cuties in Japan. In L. Skov & B. Moeran (Eds.), *Women, Media, and Consumption in Japan.* Honolulu: University of Hawai'i Press. http://basic1.easily.co.uk/04F022/036051/Cuties.html.
24. Yomota, I. (2006). *Kawaii Ron (Kawaii Theory).* Tokyo: Chikuma Shobo. (in Japanese).
25. Progressive Japanese-English Dictionary. Tokyo: Shogakukan. https://dictionary.goo.ne.jp/je/.
26. Endo, K. (2017). Column of Chapter 2, The historical genealogy of aesthetics "Kawaii". In Ohkura, M. (Ed.), *Kawaii Kogaku (Kawaii Engineering)* (pp. 9–12). Tokyo: Asakura Shoten. (in Japanese).
27. The United Nations: Sustainable Development Goals. https://sustainabledevelopment.un.org/sdgs.
28. Cheok, A. D., et al. (2008). Designing cute interactive media. *Innovation, 8*(3), 8–9.

29. Kurosu, M., et al. (2017). Chapter 2, Cuteness in Japan. In *Cuteness engineering -designing adorable products and services-* (pp. 33–61). Heidelberg: Springer (2017).
30. Shonagon, S. (1991). *The Pillow Book of Sei Shonagon* (translated and edited by Morris, I.). New York: Columbia University Press.
31. Geijutsu Shincho, (2011). *"Kawaii" in Japan, Geijutsu Shincho, September Issue of 2011.* Tokyo: Shincho-sha. (in Japanese).
32. Fuchu-City Art Museum, (2013). *Cute Edo Paintings.* Tokyo: Kyuryudo. (in Japanese).

Part II
Systematic Study on Kawaii Engineering

Chapter 2
Systematic Measurement and Evaluations for Kawaii Products

Michiko Ohkura and Tetsuro Aoto

Abstract This chapter introduces the systematic study for kawaii artificial products using questionnaires. The physical attributes employed are color, shape, material perception (visual texture, tactile texture) and sound.

Keywords Kawaii · Industrial product · Color · Shape · Hue · Brightness · Saturation · Visual analog scale · Visual texture · Tactile texture · Material perception · Onomatopia · Sound · Timbre · Loudness · Pitch

2.1 Introduction

Mainly based on the Refs. [1–7], this chapter introduces the systematic study for kawaii artificial products using questionnaires. Based on the background described in Chap. 1, we began our research on kawaii engineering. This chapter introduces the systematic study for kawaii products using questionnaires. The physical attributes employed are color, shape, material perception (visual texture, tactile texture) and sound. Size will be described in Chap. 5.

2.2 Comparison of Kawaii Colors and Shapes on a Two-Dimensional Plane

To make a new step to clarify the attributes of kawaii products, we performed an experiment for colors and shapes on a two-dimensional plane [1].

M. Ohkura (✉) · T. Aoto
Shibaura Institute of Technology, Tokyo, Japan
e-mail: ohkura@sic.shibaura-it.ac.jp

© Springer Nature Singapore Pte Ltd. 2019
M. Ohkura (ed.), *Kawaii Engineering*, Springer Series on Cultural
Computing, https://doi.org/10.1007/978-981-13-7964-2_2

Ten basic hues from the Munsell color system (red, yellow-red, yellow, yellow-green, green, blue-green, blue, blue-purple, purple, and red-purple) with the addition of white and black samples were presented to volunteers, who comprised 20 female and 20 male students in their 20s. First, they were asked to choose the most kawaii color from 12 candidates on a sheet shown in Fig. 2.1. If they could not choose a kawaii color, the answer was "no color." Then they were asked to choose the most kawaii shape from the 12 candidates shown in Fig. 2.2. If they could not choose a kawaii shape, the answer was "no shape." We employed the 12 basic shapes of the Adobe Photoshop.

Finally, they were asked to make the most kawaii combination of a color and a shape using the same color sheet shown in Fig. 2.1 and a sheet with the cutout shapes shown in Fig. 2.3. If they could not choose a kawaii combination, the answer was "no combination."

Figure 2.4 shows the experimental results for the first question. Figure 2.5 shows the experimental results for the second question. In both figures, the vertical axes show the number of participants who chose the color or the shape in the horizontal axes.

Fig. 2.1 Sheet of colors

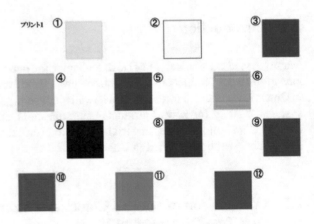

Fig. 2.2 Sheet of shapes

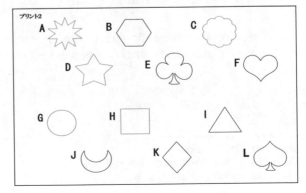

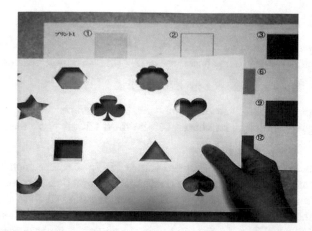

Fig. 2.3 Sheets for combination

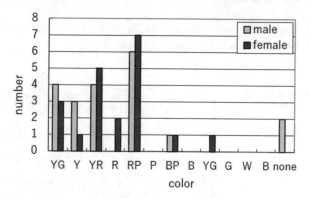

Fig. 2.4 Result of choosing most kawaii color

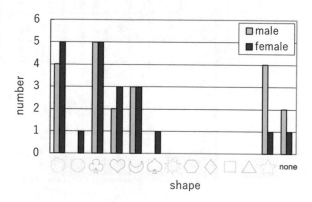

Fig. 2.5 Result of choosing most kawaii shape

The following were obtained from Fig. 2.4:

- All participants except two could choose a kawaii shape.
- Warmer colors tended to be chosen for the most kawaii more than colder colors.

The followings were obtained from Fig. 2.5:

- All participants except three could choose a kawaii shape.
- Curved shapes tended to be chosen for the most kawaii more than shapes with straight lines.

For the final question, the results are shown in Fig. 2.6, in which each number shows the number of participants who chose that combination of color and shape.

The color and shape combination chosen as the most kawaii were not necessarily the same as the results of the most kawaii color and the most kawaii shape chosen by each participant. However, the tendencies to choose warmer colors and curved shapes as the most kawaii did not change. Thus, these tendencies are considered consistent.

This simple experiment reconfirmed that the value of kawaii for artificial objects is acceptable to Japanese males and females in their 20s.

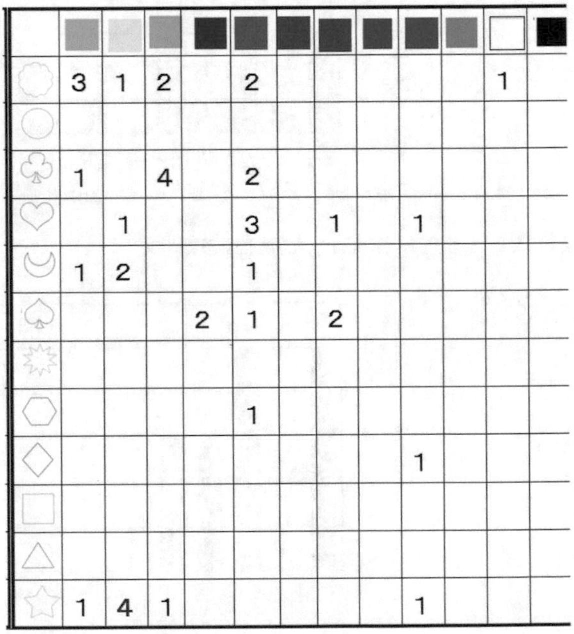

Fig. 2.6 Result of choosing most kawaii combination

Shape											
cloud	3	1	2		2					1	
oval											
club	1		4		2						
heart		1			3	1	1				
crescent	1	2			1						
spade			2	1	2						
burst											
hexagon					1						
diamond									1		
square											
triangle											
star	1	4	1						1		

2.3   Comparison of Kawaii Colors and Shapes in Three-Dimensional Space

Since the final goal of our research is to construct kawaii products, we extended our previous study for 2-dimensional objects and performed a simple experiment employing 3-dimensional objects in a 3-dimensional environment [1].

We employed a virtual environmental system constructed and used for our various previous researches [8–10]. It is shown in Fig. 2.7.

Six basic objects from 3ds Max were chosen as candidates: box, pyramid, cube, cylinder, tube, and torus. Red, blue, and green were chosen as the basic colors. Participants were presented a set of 6 objects of the same color from various viewpoints and asked to choose the most kawaii object and to explain their choice. This procedure was repeated 3 times for 3 colors. Examples of the presented sets of objects are shown in Fig. 2.8. Finally, presented with the 3 chosen objects for different colors all together, participants were again asked to choose the most kawaii object and to explain their choice.

Experiments were conducted with 6 female and 6 male students in their 20s. The temperature and humidity of the experimental room ranged from 23.3 to 24.2 °C, and 49–51%, respectively.

Figure 2.9 shows the results for each color. The vertical axis shows the number of participants who chose the object in the horizontal axis. The following conclusions were obtained:

- For blue objects, such round objects as cubes, cylinders and tori were chosen as the most kawaii.
- For red objects, most participants judged that no objects were kawaii.
- For green objects, 1/3 of the participants chose straight-lined objects, 1/3 chose round objects, and 1/3 judged none as the most kawaii object.
- The most kawaii shape of an object depends on its color.
- Differences exist by gender.

Table 2.1 shows the results of the most kawaii combination. The numerator and the denominator respectively denote the number of male and female participants who chose that combination. The following conclusions were obtained from this table.

Fig. 2.7 Experimental setup [8]

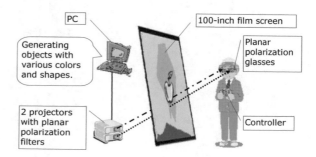

Fig. 2.8 Examples of presented 3D objects

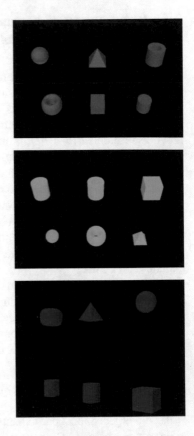

Fig. 2.9 Results of choosing 3D object

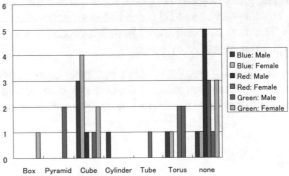

Table 2.1 Results for final combination

Male/female	Blue	Red	Green
Box			0/1
Pyramid			
Cube	3/1		1/2
Cylinder	1/0		
Tube	1/0		
Torus		0/2	
None			

- Most participants chose round objects as the most kawaii.
- Most male participants chose blue objects as the most kawaii, but female participants didn't show a color preference.

This experiment also reconfirmed that the value of kawaii for artificial objects is acceptable for Japanese males and females in their 20s. However, differences exist between the results of the first and the second experiments.

2.4   Discussion of the Two Experiments

The results of the two experiments (Sects. 2.2 and 2.3) reconfirmed that the value of kawaii for artificial objects is acceptable for young Japanese males and females. Moreover, comparing both experiments, following results are identical:

- All participants could choose a most kawaii combination of color and shape.
- The choice of the most kawaii shape varied by color.
- The most kawaii object has a curved shape.

On the other hand, differences exist by color. Warmer colors tended to be chosen as the most kawaii in the first experiment, while most male participants chose blue objects as the most kawaii in the second experiment. It is uncertain whether this was caused by the differences of the 2-dimensional plane shapes in the first experiment and the 3-dimensional virtual objects in the second experiment. Additional detailed experiments remain as future work. The next session describes the experiment.

2.5   Detailed Experiment to Compare the Kawaii Colors Between 2D and 3D Objects

2.5.1   Overview

To solve the discrepancy in kawaii color between the results obtained from the experiment using sheets of paper and the results obtained from the experiment using three-dimensional virtual space, we planned another experiment to compare both two-dimensional objects and three-dimensional objects using the same experimental equipment [2].

We employed a virtual environmental system constructed and used for our various previous studies (Sect. 2.3), shown in Fig. 2.7. For three-dimensional space, the images for both eyes had certain differences for visual disparities, and usual planar polarization glasses were used, that is, the polarized filter for the left eye had vertical stripes and the polarized filter for the right eye had horizontal stripes. On the other hand, the images for both eyes were the same, and both polarized filters for both eyes had the vertical stripes in common for the two-dimensional plane.

As for the shapes for three-dimensional objects, the same candidates used for the previous experiments (Sect. 2.3) were employed. On the other hand, square, triangle, circle, rectangle, and torus were employed as the candidates of shapes for two-dimensional objects.

As the basic colors, red, blue, and green were chosen, which are the same used in the previous experiment (Sect. 2.3).

2.5.2   Experimental Method

The procedure was similar to the previous experiment (Sect. 2.3).

Participants were presented a set of 6 three-dimensional objects of the same color from various viewpoints in three-dimensional space and asked to choose the most kawaii object and to explain their choice. Then, presented with the 3 chosen objects for different colors all together, participants were again asked to choose the most kawaii object of the three and to explain their choice.

The above procedures were repeated for 5 two-dimensional objects in the two-dimensional plane in the same manner. Examples of the presented sets of objects are shown in Fig. 2.10.

The order of colors was random, and the order of presenting three-dimensional objects and two-dimensional objects was counter-balanced.

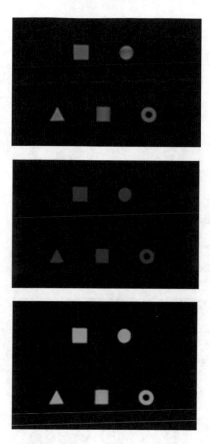

Fig. 2.10 Examples of presented 2D objects

2.5.3 Experimental Results

Experiments were performed with 6 female and 6 male students in their 20s. The temperature and humidity of the experimental room ranged from 23.7 to 24.5 °C, and 46–56%, respectively.

Figure 2.11 shows the results of three-dimensional (3D) objects for each color, and Fig. 2.12 shows the results of two-dimensional (2D) objects for each color. In both figures, the vertical axes show the number of participants who chose the object in the horizontal axis. From these figures, the following conclusions were obtained.

- For 3D blue objects, such curve-shaped objects as cubes and tori were chosen as the most kawaii.
- For 3D red objects, such curve-shaped objects as cubes, cylinders and tori were chosen as the most kawaii.
- For 3D green objects, most participants chose such curve-shaped objects as cubes and tori as the most kawaii.

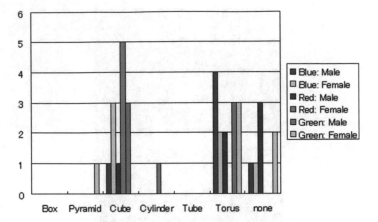

Fig. 2.11   Results of choosing most kawaii object in 3D

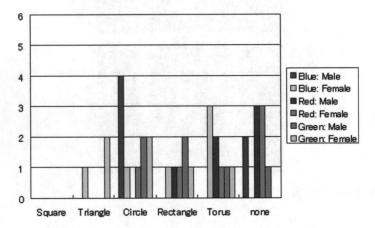

Fig. 2.12   Results of choosing most kawaii object in 2D

- The most kawaii shape of a 3D object depends on its color.
- For 2D blue objects, such curve-shaped objects as circles and tori were chosen as the most kawaii.
- For 2D red objects, half of the participants judged that neither was kawaii.
- For 2D green objects, almost half of participants chose straight-lined objects such as triangles and rectangles, and the rest chose curve-shaped objects such as cubes and tori as the most kawaii.
- The most kawaii shape of a 2D object also depends on its color.

Tables 2.2 and 2.3 show the results of the most kawaii combination for 3D objects and 2D objects respectively. The numerator and the denominator respectively denote

Table 2.2 Results for final combination in 3D

Males/females	Blue	Red	Green
Box			
Pyramid			
Cube	1/2	0/2	2/0
Cylinder			
Tube			
Torus	1/1		2/1
None			

Table 2.3 Results for final combination in 2D

Males/females	Blue	Red	Green
Square			
Triangle	0/1		
Circle	3/0	0/1	0/1
Rectangle	0/1		2/0
Torus	0/2	1/0	
Square	1/1		2/1
None			

the number of male and female participants who chose those combinations. The following was obtained from these tables.

- All participants could choose the most kawaii combination of color and shape for 3D objects and 2D objects.
- All participants chose curve-shaped objects as the most kawaii among 3D objects.
- Many but not all participants chose curve-shaped objects as the most kawaii among 2D objects.
- The choice of the most kawaii shape varied by color for both 3D objects and 2D objects.
- The color chosen least as the most kawaii combination was red for both 3D and 2D objects.

2.5.4 Discussion and Summary

Comparing the results for 3D objects and 2D objects, kawaii preference for curved shape and for blue and green colors are in common. As for the difference between the two, the number of participants who chose straight-line-shaped objects as the most kawaii was larger for 2D objects than 3D objects.

Compared the results for 3D objects with the previous results for 3D objects (Sect. 2.3), kawaii preference in curved shape and kawaii preference of blue and green in color are in common.

On the other hand, the results for 2D objects have some similarities and some differences compared with the previous results using sheets of paper (Sect. 2.2). That is, kawaii preference for curved shape is in common, whereas, kawaii preference for color is different: many participants chose 2D blue or green objects as the most kawaii, which differs the previous results that warmer colors tended to be chosen as the most kawaii (Sect. 2.2).

As described in the previous section (Sect. 2.4), it was uncertain whether the difference in the most kawaii color between 3D objects in 3D space and 2D objects on a sheet of paper was caused by the differences between the dimensions of the objects. However, the results of this experiment clarified that the difference was not caused by the dimension of the objects. Thus, the reason for the difference remains unclear. Additional detailed experiments remain as future work.

2.6   Detailed Experiment on Kawaii Colors of Virtual Objects

2.6.1   Background

There exists a discrepancy in kawaii colors between the experimental results of the previous sections (Sects. 2.2 and 2.3). That is, warmer colors tended to be chosen as the most kawaii in one experiment, while most male participants chose blue objects as the most kawaii in another experiment. By performing the experiment in Sect. 2.5, the discrepancy was not solved. Thus, to solve this discrepancy, we planned this new experiment [3].

2.6.2   Experimental Setup

We employed a 22-inch 2D/3D compatible LCD monitor by Zalman to show virtual objects. Participants wear polarized glasses to watch the objects stereoscopically.

As for the kawaii shape of objects, a torus was employed based on the results of our previous studies. To select the candidates of kawaii colors, we used the Muncell Color System (MCS). Color has three elements: hue, saturation, and brightness. As for hues, we employed five basic hues based on MCS. For each hue, three connected values of saturation and three connected values of brightness were selected as shown in Fig. 2.13 and Table 2.4. Thus, the total number of kawaii color candidates was 45. The background color was set as gray with a saturation of five.

Fig. 2.13 Selected candidates for red

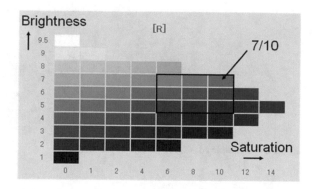

Table 2.4 Selected candidates for five basic hues

(R)			
B/S	6	8	10
7	7/6	7/8	7/10
6	6/6	6/8	6/10
5	5/6	5/8	5/10
(Y)			
B/S	6	8	10
8	8/6	8/8	8/10
7	7/6	7/8	7/10
6	6/6	6/8	6/10
(G)			
B/S	4	6	8
7	7/4	7/6	7/8
6	6/4	6/6	6/8
5	5/4	5/6	5/8
(B) (P)			
B/S	2	4	6
7	7/2	7/4	7/6
6	6/2	6/4	6/6
5	5/2	5/4	5/6

2.6.3   Experimental Method

Participants were presented three objects out of nine of the same hue and asked to choose the most kawaii object from the three. After this procedure was repeated three times to present all candidates of one of the basic five hues, the selected objects were shown again for the final determination of the most kawaii object for the hue.

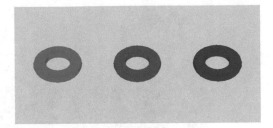

Fig. 2.14 Example of a set of objects on the screen

After the above procedures were repeated five times for all basic hues, the five selected objects were shown again for the final determination of the most kawaii object for all.

The orders of showing three objects out of nine for each hue and the orders of showing five hues were both set to be random. An example of the presented sets of objects is shown in Fig. 2.14.

2.6.4 Experimental Results

Experiments were performed with 12 female and 12 male students in their 20s with normal or normally-corrected eyesight. The temperature and humidity of the experimental room ranged from 25 to 26 °C, and 50–64%, respectively.

Figure 2.15 shows an example of the results for each hue, where the vertical axis shows the number of participants who chose the object as most kawaii with the pair of brightness and saturation shown in the horizontal axis. The results of the analysis of variance with three elements, brightness, saturation, and gender, for each hue are as follows.

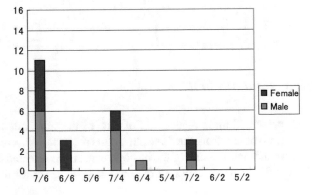

Fig. 2.15 Example of the results (blue)

Fig. 2.16 Final result of most kawaii

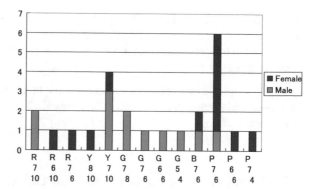

- Brightness is significantly effective in red, blue, and purple.
- Saturation is significantly effective in purple.
- Gender is not significant for all hues.

Figure 2.16 shows the final result of the most kawaii color, where the horizontal axis is the combination of hue, brightness, and saturation.

2.6.5   Discussion

From Figs. 2.15 and 2.16, the following is obtained.

- With more brightness, more participants chose the object as most kawaii for every hue.
- With more saturation, more participants chose the object as most kawaii for every hue. However, the differences were smaller than those for brightness except for yellow.
- All hues were chosen as most kawaii by at least one participant. The most chosen hue was purple, and the next was yellow. Most participants who chose purple were female, while most participants who chose green or blue were male.

These results show the importance of purple and yellow as candidates for kawaii hues. Thus, the discrepancy between the results of our previous experiments may be caused by the condition that only red, blue, and green were candidates of hues in some experiments.

2.6.6   Summary

In this study, we focused on the kawaii color of virtual objects to apply to interactive systems and industrial products in the future. From the experiments, the following findings were obtained.

- Brightness is effective for kawaii color.
- Saturation is also effective for kawaii color.
- All hues can be chosen as the most kawaii color, while purple and yellow tend to be chosen most often.

Comparison between pure color and the most bright and saturated candidates for each hue in this experiment remain for future work. The experiment will be described in the next section.

2.7   Detailed Experiment on Kawaii Colors Using Visual Analog Scale

2.7.1   Background

As the result of the previous section (Sect. 2.6), with more brightness and more saturation, more participants chose the object as most kawaii for every hue. The most chosen hue was purple for both males and females. However, since we have not studied intermediate hues based on MCS, we addressed this problem in a new experiment. This section describes the experimental results [4].

2.7.2   Experimental Set-Up

We showed the virtual objects on a 46-inch 2D/3D compatible LCD monitor from Hyundai on which participants watched stereoscopically with polarized glasses. For the kawaii shapes of objects, a torus was employed based on the results of our previous studies (Sect. 2.3). To select the candidates of kawaii colors, we used MCS. Color has three elements: hue, saturation, and brightness. We employed 5 basic hues (R, Y, G, B, and P) and 5 intermediate hues (YR, GY, BG, PB, and RP) based on MCS. For each hue, the following 4 colors were selected:

- #1 is white, which has the highest brightness and the lowest saturation.
- #2 has higher brightness and lower saturation than the base color.
- #3 is a base color with high brightness and high saturation.
- #4 is pure color, which has lower brightness than the base color and the highest saturation.

These colors for each hue were selected (Fig. 2.17). Because #1 for each hue is the same color, the total number of kawaii color candidates was 31. The background color was gray.

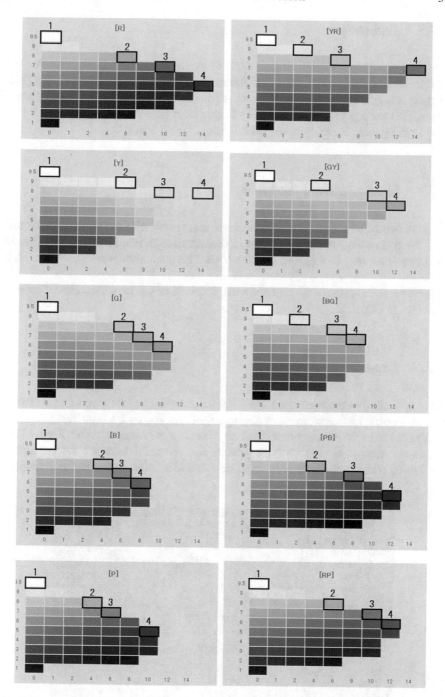

Fig. 2.17 Selected colors for each hue

2.7.3 Evaluation

We used the Visual Analog Scale (VAS) to evaluate the kawaii degrees. For the evaluation of pain severity and relief, the method commonly used is the VAS [11]. Participants arbitrarily marked 200-mm segments. The left side line doesn't seem kawaii, but the right side line does. The length from the left side to the mark put on segments by participants is converted into scores from 0 to 100.

2.7.4 Experimental Procedure

First, the color blindness of the participants was tested with the Ishihara color test [12]. Next, they were shown four colors of the same hue and simultaneously evaluated the kawaii degrees of the four colors with VAS. This evaluation was repeated for each hue.

The ten hues were shown randomly. An example of the presented sets of objects is shown in Fig. 2.18.

2.7.5 Experimental Results

The experiments were performed with 10 female and 10 male students in their twenties with normal or normally-corrected eyesight checked by Ishihara Test [12].

The kawaii scores were normalized on a basis of the score of white. Figure 2.19 shows an example of the results by gender. The vertical axis shows the average of the kawaii scores with the pair of brightness and saturation shown in the horizontal

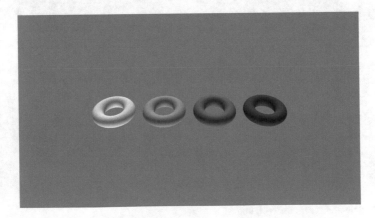

Fig. 2.18 Example of a set of objects

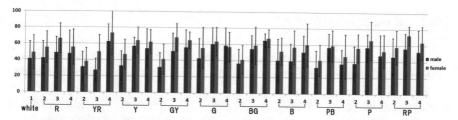

Fig. 2.19   Averages of kawaii degrees of each color

axis. The error bars indicate the standard deviations. We performed ANOVA with three factors: hue, brightness/saturation, and gender.

The following were obtained from ANOVA:

- The main effects of gender, hue, and brightness/saturation are significant.
- The interaction effect between hue and brightness/saturation is significant.

The following were obtained from Fig. 2.19:

- Color G#3 was evaluated high by males.
- Colors RP#3 and GY#3 were evaluated high by females.
- Colors YR#4 and BG#4 were evaluated high by both genders.
- Color #2 of each hue was evaluated low by both genders.
- Colors with intermediate hues based on MCS got relatively high scores.

2.7.6   Summary

We focused on the kawaii color of virtual objects to apply to future interactive systems and industrial products and obtained the following findings:

- The kawaii scores for colors differ by gender, hue, and brightness.
- The combinations of hue and brightness/saturation are important for evaluations of the kawaii colors.
- Pure yellow red and pure blue green are evaluated high by both genders and are felt to be kawaii.

2.8   Material Perception for Visual Texture

This section is described based on the Ref. [5].

2.8.1 Background Survey

The first author lectured on kawaii research and asked the undergraduate students to list five kawaii products or living creatures and to explain their reasons in November 2010. The listed goods were classified into three categories: living creatures, artificial products, and cartoon characters (Table 2.5). We morphologically analyzed their explanations for the kawaii feelings and made histograms of the morphemes. Figure 2.20 shows histograms of the physical attributes for the three categories. We confirmed that shape, color, and size were important attributes for feeling kawaii. Our next target is material perception.

2.8.2 Kawaii Feeling on Visual Material Perception for Virtual Objects

This section is also described based on the Ref. [5].

Table 2.5 Classification of Kawaii goods

Class	Examples
Living creatures (LC)	Ant, dog, cat, penguin, bird, sparrow, frog, child, girl, baby, hamster, panda, chick
Artificial products (AP)	iPod, mobile phone, strap, round button, cake, necklace, stuffed animal
Cartoon characters	Doraemon, Pikachu, Winnie the Pooh, Donald Duck, Stitch, Moomin

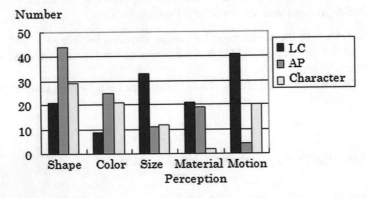

Fig. 2.20 Histogram of physical attributes

Fig. 2.21 Nine displayed objects

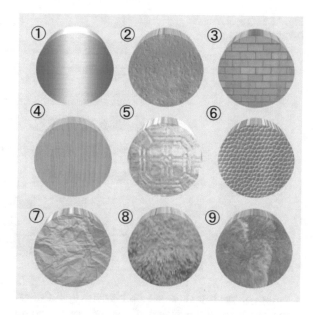

2.8.2.1   Method

We employed a 46-inch-LCD monitor and polarized glasses to stereoscopically show virtual objects for the experimental setup. Based on the results of our previous experiments described in Chap. 2, we employed a cylinder for the shape of the objects, and pink for their color.

We employed 9 textures (Fig. 2.21). Participants were randomly presented 9 objects one by one for 20s to evaluate their kawaii degree on a 7-point Likert Scale, where −3 is extremely non-kawaii, 0 is neutral, and +3 is extremely kawaii. Participants also chose the most kawaii from the 9 objects.

2.8.2.2   Results and Discussion

We performed our experiments with 9 female and 9 male students in their 20s. Figure 2.22 shows the results for each texture, where the vertical axis shows the number of texture and the horizontal axis shows the ratio of each scale.

- Each texture got both scores of the kawaii group such as +3, +2, and +1 and the non-kawaii group such as −3, −2, and −1.
- Textures #9, #8, #3, and #4 have relatively high scores.
- Textures #2, and #5 have relatively low scores.

We obtained the following results:

Fig. 2.22 Questionnaire results for each texture

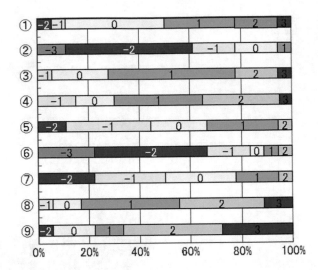

- Although each texture has both positive and negative scores, a large difference exists among the averages of the textures; a product's texture affects its kawaii degree.
- Textures evoking words related to such tactile sensation as soft, furry, and tangible are generally kawaii.
- Textures #2, and #5 have relatively low scores.

2.9 Experiment of Kawaii Feeling in Material Perception for Real Tactile Materials

From the results of the above experiment, we performed the experiment using real tactile materials. This section is described based on the Ref. [6].

2.9.1 Method

We employed 109 materials with different tactile sensations [13]. The features of those materials were that they were linked to Japanese onomatopoeias such as "mokomoko" and "zarazara" [13]. Examples of tactile materials are shown in Fig. 2.23. Blindfolded participants were shown paired materials one by one and asked to determine which is more kawaii. The answer of participant was "Right," "Left," or "Same." All materials were ranked by a quick sort algorithm (Fig. 2.24).

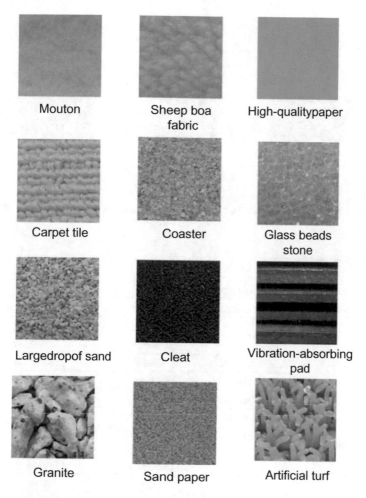

Mouton

Sheep boa fabric

High-qualitypaper

Carpet tile

Coaster

Glass beads stone

Largedropof sand

Cleat

Vibration-absorbing pad

Granite

Sand paper

Artificial turf

Fig. 2.23 Examples of tactile materials

2.9.2 *Results and Discussion*

We performed experiments with 2 female and 2 male participants. It took 2–3 h for each experiment and learned that 4 materials were deemed kawaii the first 2 times by all participants and 8 materials by 3 out of 4 participants. On the other hand, no materials were deemed not-kawaii the first 2 times by all participants and 4 materials by 3 out of 4 participants. Table 2.6 shows 4 kawaii and 4 not-kawaii materials with linked onomatopoeias. These results indicate that we can define kawaii (and not-kawaii) tactile materials using the materials we employed.

From the comparison results for each participant, we ranked the kawaii degree of each material, where the ranks were identical when the materials were judged the

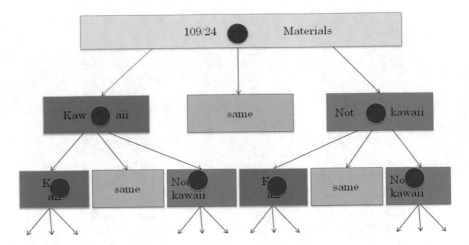

Fig. 2.24 Quick sort algorithm

Table 2.6 Outline of kawaii and not-kawaii tactile materials

Kawaii/not-kawaii	Tactile materials	Onomatopoeia 1	Onomatopoeia 2
Kawaii	Yarn	Jashijashi	Wasawasa
	Cotton	Fukafuka	Mokomoko
	Sheep boa fabric	Pofupofu	Mofumofu
	Cotton cloth	Fusafusa	Mosamosa
Not-kawaii	Large drop of sand	Jarijari	
	Granite stone	Gorogoro	Zaguzagu
	Vibration-absorbing pad	Kunikuni	Pokopoko
	Sand paper	Jusajusa	Jorijori

same by the participants. Then we averaged the kawaii ranks of all the participants. We defined the materials from the top and the 20th as the kawaii group, and materials from 90th to the bottom as the non-kawaii group. By comparing these two groups, we obtained the following:

- Kawaii group has such features as bushy, fluffy, soft, smooth, and elastic. On the other hand, non-kawaii group has such features as crumbly, hard, and rough.
- The onomatopoeias of the kawaii group have such consonants as /f/, /m/, and /p/. On the other hand, those of the non-kawaii group have /z/, /j/, and /g/.

These results resemble those of the experiment for virtual visual objects and suggest that common features of kawaii feeling exist in material perception.

2.10   New Experiment for Kawaii Feelings of Tactile Material Perception

This section is described based on the Ref. [6].

2.10.1   Method

We selected 24 materials out of 109 materials employed in the previous experiment described in Sect. 2.9 and their corresponding onomatopoeia as follows:

- The 24 materials are spread from the most kawaii to the least kawaii.
- Each selected material has /a/, /u/ or /o/ as the vowel of the first syllable of its corresponded onomatopoeia.
- All consonants such as /p/, /m/, have at least 2 corresponding materials which have those consonants in their first syllables of corresponding onomatopoeia.

Blindfolded participants were shown paired materials one by one and asked to determine which is more kawaii, as the same way as the previous experiment.

2.10.2   Results

We performed experiments with 30 participants, 10 females in their 20s, 10 males in their 20s, 5 females in their 40s or 50s, and 5 males in their 40s or 50s. Averaged orders for all materials of four participant's groups are shown in Fig. 2.25, where the vertical axis shows the averaged kawaii order and the horizontal axis shows the tactile materials. The material table in averaged kawaii order is shown in Table 2.7. The correlation coefficients for four participant's groups are shown in Table 2.8. From these results, we obtained the following:

- The averaged orders for four groups have strong positive correlations, which mean that there are no differences between genders and generations.
- The most kawaii materials in the average are Mouton, Cotton, Sheep boa fabric, and Carpet tile regardless of genders and generations.
- The onomatopoeias of the most kawaii materials have such consonants as /f/, and /m/. On the other hand, those of the least kawaii materials have /z/, /j/, and /g/.

In addition, the most kawaii materials have physical features as bushy, fluffy, soft, and "like animal hair." This tendency is similar to both previous experiments for visual and tactile material perceptions.

Table 2.7 Material table
with averaged kawaii order
with corresponding
onomatopoeia

Averaged order	Tactile material	Onomatopoeia 1	Onomatopoeia 2
1	Mouton	Fusafusa	Mofumofu
2	Cotton	Fukafuka	Mokomoko
3	Sheep boa fabric	Pofupofu	Mofumofu
4	Carpet tile	Fusafusa	Mowamowa
5	High-quality paper	Sawasawa	Subesube
6	Refrigerant	Bunyabunya	Bnyubnyu
7	Glass beads stone	Shakashaka	Tsubutsubu
8	Coaster	Kasakasa	Shurushuru
9	Slime	Zubuzubu	Zubozubo
10	Cardboard	Kasakasa	Kasukasu
11	Glossy paper	Subesube	Tsurutsuru
12	Aluminum board	Tsurutsuru	Subesube
13	Dish	Tsurutsuru	Gochigochi
14	Aluminum foil	Kasakasa	Gashigashi
15	Large drop of sand	Jarijari	–
16	Silicon rubber	Kunyukunyu	Nyuninyuni
17	Non-slip insole	Zarazara	Butsubutsu
18	Wet cloth	Gushogusho	–
19	Water	Jabujabu	–
20	Vibration-absorbing pad	Kunikuni	Pokopoko
21	Soaked cloth	Guchogucho	–
22	Granite stone	Gorogoro	Zaguzagu
23	Sand paper	Jusajusa	Jorijori
24	Artificial turf	Zakuzaku	Jogijogi

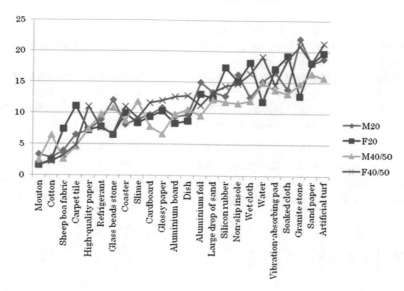

Fig. 2.25 Averaged orders of all materials

Table 2.8 Correlation coefficients

	Male 20s	Female 20s	Male 40s or 50s	Female 40s or 50s
Male 20s	–			
Female 20s	0.78	–		
Male 40s or 50s	0.88	0.73	–	
Female 40s or 50s	0.88	0.82	0.87	–

2.11   Discussion for the Experiments of Material Perception

We performed three experiments about the kawaii feelings of material perception. The first experiment (Sect. 2.8) was on kawaii feelings for various visual textures of virtual objects, and the second and the third experiments (Sects. 2.9 and 2.10) were on kawaii feelings for various tactile textures for actual materials. We can confirm that common features of kawaii feelings exist in material perception.

As the results of our previous studies for visual attributes of objects (Sects. 2.2–2.7), we obtained the following:

- We can evaluate kawaii shapes, and the round shapes are judged more kawaii regardless of genders.
- We can evaluate kawaii colors, and there exist a certain difference for kawaii colors between genders.

From the experimental results for tactile attributes of objects described in Sects. 2.9 and 2.10, we obtained the following:

- We can evaluate kawaii tactile feelings.
- There is no difference between genders and generations for kawaii tactile sensations of tactile material perception.
- The onomatopoeias of the most kawaii materials have such consonants as /f/, and /m/. On the other hand, those of the least kawaii materials have /z/, /j/, and /g/.
- The physical features of kawaii tactile materials are similar to those evoked from kawaii textures, which are bushy, fluffy, soft, and "like animal hair."

For manufacturing process, shapes and tactile features of materials for an industrial product are more difficult to prepare various types compared with the preparation of various colors. Therefore, it is a useful finding to clarify the tendencies of attributes of materials with kawaii tactile feeling in common between genders and generations.

2.12   Kawaii Sound

2.12.1   Background

This section is described based on the Ref. [7]. We experimentally evaluated kawaii visual and tactile stimuli as we already described in previous sections in this chapter. However, since we did not research kawaii auditory stimuli, we began to perform experiments to clarify the kawaii attributes of sound by focusing on its three fundamental elements: pitch, loudness, and timbre. To clarify sound's kawaii attributes, we individually experimented with these 3 sound elements.

2.12.2   Experiment on Kawaii Timbre

To clarify whether sound's timbre affects its kawaii feeling, we experimented with 16 different timbres that consisted of 15 musical/orchestral instruments and a music box. Our participants evaluated the kawaii degree for each timbre. The tempo of the sounds was set to 120 BPM and the pitch was set to 440 Hz ("la" in the musical scale). Each sound was repeated twice with a two-beat pause. The sound pressure was set to 38 ± 1.5 dBA. Under these conditions, participants randomly listened to 16 sounds with different timbres through headphones. We used the Visual Analog Scale [11] for the kawaii evaluation.

Six males in their 20s participated in the experiment. The averages with the standard deviations of their evaluation score are shown in Fig. 2.26. A one-factor ANOVA showed a significant main effect of sound timbre at a 1% level for the kawaii scores. This result suggests that a sound's timbre affects its kawaii evaluation. In addition, multiple comparison results revealed (between pairs of instruments) the following significant differences of the averaged kawaii scores ($p < 0.05$):

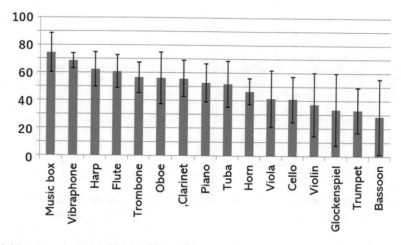

Fig. 2.26 Averages with standard deviations of kawaii scores for each timbre

- Between music box and violin
- Between music box and bassoon
- Between music box and trumpet
- Between music box and glockenspiel
- Between vibraphone and bassoon
- Between vibraphone and trumpet
- Between vibraphone and glockenspiel

These results identify significant differences of the averaged kawaii scores between the highest group (music box and vibraphone) and the lowest group (bassoon, trumpet, and glockenspiel). In addition, because the standard deviations of the lowest group timbres are relatively large, the kawaii scores of the non-kawaii timbres have larger individual differences.

Based on these results, we employed music box and vibraphone for the timbre in our experiments. The frequency analysis results of these two instruments are shown in Fig. 2.27; each timbre has a strong fundamental. On the other hand, since the timbres of the lowest group have relatively stronger components than other fundamentals, a timbre with a stronger fundamental increases the kawaii feelings.

2.12.3   Experiment on a Sound's Kawaii Loudness

We experimentally clarified the effect of sound loudness on kawaii perceptions with 12 kinds of sounds under the following conditions:

(1) Two timbres: music box and vibraphone
(2) Three sound pitches: 440, 880, and 1760 Hz
(3) Two initial sound loudness values: 31 and 51 dBA

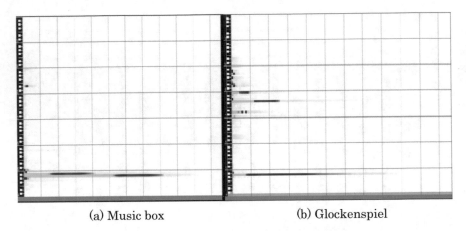

| (a) Music box | (b) Glockenspiel |

Fig. 2.27 Examples of the frequency analysis results (horizontal axis: power, vertical axis: frequency)

Participants used a numeric keypad in the experiment. After pressing 5 on it, they heard a sound for two seconds. Then, they adjusted the loudness by pressing the keyboard's numeric key and heard the different loudness sound for two seconds. The loudness changed by approx. 0.4 dBA each time the numeric key was pressed. This procedure was repeated until they chose the sound with the most kawaii loudness. The background noise was approx. 30.0 dBA.

Experiments were performed with 16 students (8 males and 8 females) in their 20s. Figure 2.28 shows the averages with the standard deviations of the most kawaii loudness of each sound. The two-factor ANOVA result for kawaii loudness showed a significant main effect of sound pitch at a 1% level and a significant main effect of gender at a 5% level. As a result of multiple comparisons for sound pitch, we identified a significant difference between 440 and 1760 Hz. The kawaii loudness of 1760 Hz was significantly larger than that of 440 Hz. This result suggests that a sound's most kawaii loudness differs among various sound pitches, and a sound with a higher frequency has greater loudness for the most kawaii sound. The most kawaii loudness was also larger for females than males.

2.12.4 Experiment on a Sound's Kawaii Pitch

We experimentally investigated the influence of a sound's pitch on its kawaii evaluation. In the experiment, we presented the following four sounds:

(1) Timbre: music box, initial value of sound pitch: E4 (329.63 Hz);
(2) Timbre: music box, initial value of sound pitch: A 6 (1760.0 Hz);
(3) Timbre: vibraphone, initial value of sound pitch: E4 (329.63 Hz);
(4) Timbre: vibraphone, initial value of sound pitch: A6 (1760.0 Hz).

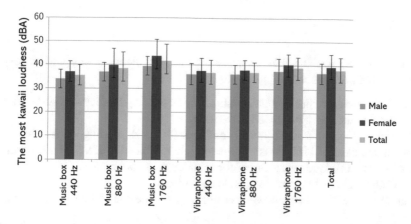

Fig. 2.28 Averages with the standard deviations of the most kawaii loudness for each sound

Participants used a numeric keypad in the experiment. After pressing the enter key, they heard a sound for two seconds. After adjusting the sound's pitch by pressing the numeric key, they heard a sound with different pitch. The sound's pitch shifted one scale each time the numeric key was pressed. We repeated this procedure until they chose a sound's most kawaii pitch from among 37 candidates from C4 (261.63 Hz) to C7 (2093 Hz) for sounds (1) to (4). At the experiment's end, each participant answered questionnaires.

Experiments were performed with 16 students (eight males and eight females) in their 20s. Table 2.9 shows the selection results. A sound pitch of C7 (2093 Hz) was selected most often as the most kawaii pitch. However, since C7 is the highest sound pitch among the presented sounds, future work must determine whether an even higher pitch might be more kawaii.

Table 2.9 Numbers of participants who chose each pitch as most kawaii

Pitch	C4	C#4	D4	D#4	E4	F4	F#4	G4	G#4	A4	A#4	B4
Number	0	0	0	0	1	6	0	0	0	2	2	4
Pitch	C5	C#5	D5	D#5	E5	F5	F#5	G5	G#5	A5	A#5	B5
Number	5	0	0	0	0	2	0	2	1	1	0	0
Pitch	C6	C#6	D6	D#6	E6	F6	F#6	G6	G#6	A6	A#6	B6	C7
Number	3	0	1	1	0	0	2	4	3	5	3	1	15

2.12.5 Summary

We clarified the kawaii attributes using 3 fundamental elements of sound and experimentally obtained the following results:

- Timbre affects a sound's kawaii evaluation, suggesting that it has a strong fundamental that increases its kawaii evaluation.
- A particular sound's most kawaii loudness differs depending on its pitch. The kawaii loudness of 1760 Hz is significantly larger than that of 440 Hz. In addition, the most kawaii loudness is larger for females than males.
- C7 (2093 Hz) or higher is the most kawaii pitch.
- Future work will experiment with combinations of the three fundamental elements.

2.13 Conclusion

This chapter introduced the systematic study for kawaii products using questionnaires. The physical attributes employed were color, shape, material perception (visual texture, tactile texture), and sound.

Curved shapes such as a torus and a sphere are generally evaluated as more kawaii than straight-lined shapes. However, a discrepancy existed in kawaii colors among the experimental results. Warmer colors tended to be chosen as the most kawaii in one experiment, while in another experiment most male participants chose blue objects as the most kawaii. Thus, to solve this discrepancy, we performed a new experiment that employed the Muncell Color System (MCS), where a color has three elements: hue, saturations, and brightness. The following are the obtained results for colors of 3D objects:

- The kawaii scores for colors differ by gender, hue, brightness, and saturation on MCS.
- Brightness is effective for kawaii color. Brighter color is more kawaii.
- Saturation is also effective for kawaii color. Higher saturation color is more kawaii.
- Colors with intermediate hues get relatively high kawaii scores.
- The combinations of hue and brightness/saturation are important for evaluations of the kawaii colors.

We performed three experiments about the kawaii feelings of material perception. The first experiment (Sect. 2.8) was on kawaii feelings for various visual textures of virtual objects, and the second and the third experiments (Sects. 2.9 and 2.10) were on kawaii feelings for various tactile textures for actual materials. We could confirm that common features of kawaii feelings exist in material perception and there is no difference between genders and generations for kawaii tactile sensations of tactile material perception. For manufacturing process, it is a useful finding to clarify the tendencies of attributes of materials with kawaii tactile feeling in common between genders and generations.

In addition, the experimental results described in the Sect. 2.12 showed that timbre, loudness, and pitch of sound affect kawaii feeling of the sound.

Acknowledgements This research was partly supported by the Grant-in-Aid for Scientific Research (C) (No.17500150), (C) (No. 21500204), and (B) (No. 2628104), Japan Society for the Promotion of Science, and by the SIT Research Promotion Funds. We thank Prof. M. Sakamoto of The University of Electro-Communications for supplying tactile materials with related onomatopoeia for Sects. 2.9 and 2.10. We also thank the students of Shibaura Institute of Technology who contributed to the research and served as volunteers.

References

1. Ohkura, M., Konuma, A., Murai, S., & Aoto, T. (2008). Systematic study for "kawaii" products (The Second Report)-comparison of "kawaii" colors and shapes. In *Proceedings of SICE Annual Conference 2008, Chofu, Tokyo* (pp. 481–484).
2. Murai, S., Goto, S., Aoto, T., & Ohkura, M. (2008). Systematic study on kawaii products (The Third Report)-comparison of kawaii between 2D and 3D. In *Proceedings of Annual Conference of The Virtual Reality Society in Japan 2008, The Virtual Reality Society in Japan, Nara* (pp. 544–547). (in Japanese).
3. Ohkura, M., Goto, S., & Aoto, T. (2009). Systematic study for 'kawaii' products: Study on kawaii colors using virtual objects. In *Proceedings of the 13th International Conference on Human-Computer Interaction* (pp. 633–637). Springer.
4. Komatsu, T., & Ohkura, M. (2011). Study on evaluation of kawaii colors using visual analog scale. In *Human interface and the management of information, Part I* (pp.103–108). Berlin Heidelberg: Springer.
5. Ohkura, M., & Komatsu, T. (2013). Basic study on kawaii feeling of material perception. In *Proceedings of HCI International 2013* (pp. 585–592). Springer.
6. Ohkura, M., Osawa, S., Komatsu, T. (2013). Kawaii feeling in material perception. In *Proceedings of IASDR2013, 07C-2*.
7. Ohkura, M., & Kanno, R. (2014). Systematic study on kawaii products (The seventh report)-Basic study for kawaii sound. *IEICE Technical report, 114*(52), 389–392. (in Japanese).
8. Aoto, T., et al. (2005). Construction of virtual Toyosu campus display system. *The Research Report of Shibaura Institute of Technology, Natural Sciences and Engineering, 49*(1), 11–18. (in Japanese).
9. Ohkura, M., et al. (2005). Comparison of the impression of the space between virtual environment and real environment. In *Proceedings of the 11th International Conference on Human-Computer Interaction 2005*.
10. Ohkura, M., et al. (2007). Evaluation of comfortable spaces for women using a virtual environment. In *Proceedings of the 12th International Conference on Human-Computer Interaction 2007*.
11. Bird, S. B., & Dickson, E. W. (2001). Clinically significant changes in pain along the visual analog scale. *Annals of Emergency Medicine, 38*(6), 639–643.
12. Ishihara, S. (1916). *Test for color-blindness*. Tokyo: Handaya. (in Japanese).
13. Watanabe, J., Kano, A., Shimizu, Y., & Sakamoto, M. (2011). Relationship between judgments of comfort and phonemes of onomatopoeias in touch. *Transactions of the Virtual Reality Society of Japan, 16*(3), 367–370. (in Japanese).

Chapter 3
Affective Evaluation for Material Perception of Bead-Coated Resin Surfaces Using Visual and Tactile Sensations—Focusing on Kawaii

Michiko Ohkura, Wataru Morishita, Kazune Inoue, Ryuji Miyazaki, Ryota Horie, Masato Takahashi, Hiroko Sakurai, Takashi Kojima, Kiyotaka Yarimizu and Akira Nakahara

Abstract We performed affective evaluation experiments on both visual and tactile material perception of bead-coated resin surfaces in which the materials, the diameters, and the hardness of beads were expanded systematically. First, we performed an experiment on the similarity of the bead-coated resin surfaces to reduce number of material samples. Second, another experiment was performed on the similarity of the evaluation items to reduce their number. Third, we performed an affective evaluation experiment on selected samples with students and middle-aged participants to clarify the differences among the different physical attributes of resin surfaces and focused on kawaii. Then, we added hues, which is one of the three elements of color, and performed two preliminary experiments to select appropriate adjective pairs for affective evaluation of the effect of hues and to select the hue candidates. With 21 adjective pairs (e.g., masculine–feminine and relaxed–anxious), and seven hues (e.g., yellow-red and blue-purple), we clarified the combined effect of the tactile sensation and the hues on the affective evaluations using combinations of actual bead-coated resin and 3D models. Our analysis results suggest that hue, bead diameter, and gender are important for affective evaluation.

Keywords Visual sensation · Tactile sensation · Resin · Bead-coated resin surface · Affective evaluation · Kawaii · Hue · 3D model

M. Ohkura (✉) · W. Morishita · K. Inoue · R. Miyazaki · R. Horie
Shibaura Institute of Technology, Tokyo, Japan
e-mail: ohkura@sic.shibaura-it.ac.jp

M. Takahashi · H. Sakurai
R&D Department, DIC Corporation, Sakura-City, Japan

T. Kojima · K. Yarimizu · A. Nakahara
Dispersion Technical Department, DIC Corporation, Tokyo, Japan

© Springer Nature Singapore Pte Ltd. 2019
M. Ohkura (ed.), *Kawaii Engineering*, Springer Series on Cultural
Computing, https://doi.org/10.1007/978-981-13-7964-2_3

3.1 Introduction

We previously performed experiments for such kawaii attributes as visual textures in virtual environment and actual tactile materials (Chap. 2). However, these textures and materials were not chosen under the systematic distribution of physical parameter values. We used bead-coated resins for affective evaluations in which each bead was made from one of three resin materials of a particular size and hardness in systematic expansions.

This chapter introduces our affective evaluation results of the surfaces of these bead-coated resins by focusing on kawaii [1–3].

3.2 Experiments for Kawaii Evaluation for Bead-Coated Resin Surfaces

3.2.1 Bead-Coated Resin Samples

Bead-coated resin samples were wrapped on the sides of cans with the following characteristics:

- Resin material: PE, VI, and FP.
- Bead diameter:#1 (6 μm)–#7 (93 μm) (Table 3.1).
- Bead hardness: soft (S), medium (M), hard (H), and super-hard (SH).

Three samples with different resin materials without bead-coats were also prepared.

3.2.2 Experiment for Sample Similarity

To reduce the number of samples for an affective evaluation experiment, an experiment was performed using 30 samples: ten samples from Table 3.1 for each of three resin materials.

The pairs of samples shown in Tables 3.2 and 3.3 were presented to the participants who rated the similarity as similar (+1), not similar (−1), or neither (0). Each sample

Table 3.1 Combination of particle diameter and bead hardness

	#1	#2	#3	#4	#5	#6	#7
S				○			
M	○	○	○	○	○	○	○
H				○			
SH				○			

Table 3.2 Combination of comparison pairs for bead hardness

	S	M	H	SH
S				
M	○			
H		○		
SH	○		○	

Table 3.3 Combination of comparison pairs for particle diameter of beads and visual similarity results for PE

	#1	#2	#3	#4	#5	#6	#7
#1							
#2	0.9						
#3		1.0					
#4	0.4		0.6				
#5		0.0		0.2			
#6			0.0		1.0		
#7				0.0		0.6	

Fig. 3.1 Visual presentation

pair was presented either visually or tactilely. The presentation order was counterbalanced.

Six participants in their 20s served as volunteers. Figures 3.1 and 3.2 show experimental scenes in visual and tactile presentation, respectively. Table 3.3 shows the averaged scores of visual similarities for PE. Based on all the results, we excluded one sample of each pair whose average similarities exceeded 0.8 both visually and tactilely. The samples for the affective evaluation were determined as follows:

- PE: N, #2, #3, #4, #5, #6, #7, M, H, SH.
- VI: N, #2, #3, #4, #5, #6, #7, M.
- FP: N, #2, #3, #4, #5, #6, #7, M, H, SH,

where N means the non-bead-coated sample.

Fig. 3.2 Tactile presentation

3.2.3 Experiment for Evaluation Factor Similarity

The candidates for the factors of the affective evaluation were collected from a survey of previous research. To choose one from some similar factors, we experimentally evaluated the similarities among factors using 12 samples: N, #1, #4, and #7 of each of the three resin materials. Participants were presented pairs of samples and asked to compare each one with a standard sample, which was one of the non-bead-coated samples. A 7 Likert scale was employed for the adjective pairs, and a 5 Likert scale was employed for the adjectives. "Impossible to evaluate" was included as an option for all the adjective pairs and adjectives. Each sample pair was presented either visually or tactilely.

Six participants in their 20s served as volunteers. After excluding the adjective pairs and the adjectives with many "Impossible to evaluate" answers, we classified them by clustering analysis. Table 3.4 shows the final results, the list of the adjective pairs, and the adjectives used for the affective evaluation experiment described below.

Table 3.4 Adjective pairs and adjectives

Adjective pair	Adjective
Chic–Ordinary	Smooth
Simple–Gaudy	Damp
Masculine–Feminine	Granular
Comfortable–Uncomfortable	Sticky
Interesting–Boring	Dry
Hot–Cold	Slippery
Easy to hold–Hard to hold	Wild
Artificial–Natural	Mild
Adult–Childish	Exciting
Favorite–Hateful	Staid
	Cool
	Kawaii

3.2.4   Affective Evaluation Experiment of Bead-Coated Resins

3.2.4.1   Method

Based on the results of the above experiments mentioned in Sects. 3.2.2 and 3.2.3, we performed another experiment on the affective evaluation of the bead-coated resins of 28 samples using the adjective pairs and adjectives from Table 3.4. The samples were presented in three ways: visually, tactilely, or both. The orders of these three ways were counter-balanced between visually and tactilely, and both were always the last. The standard sample of each pair was one of the three samples without beads of the three resin materials. The evaluation procedure was the same as the experiment in the previous section.

3.2.4.2   Results and Discussion

Experiments were performed with 24 participants, 6 males in their 20s, 6 females in their 20s, 6 males in their 40s or 50s, and 6 females in their 40s or 50s.

We separately performed several analyses for the 20s groups and the 40s/50s groups. A three-way analysis of variance (ANOVA) was performed for the evaluation results of each adjective pair/adjective. Table 3.5 shows the ANOVA results of each adjective with bead hardness, the resin material of the beads, and gender as factors for the 20s. Table 3.6 shows the ANOVA results for each adjective with bead diameter, the resin material of the beads, and gender as factors for the 20s.

Table 3.5 ANOVA for bead-hardness, resin material, and gender of 20s

Adjective	Visual			Tactile			Both		
	Hardness	Material	Gender	Hardness	Material	Gender	Hardness	Material	Gender
Smooth									
Damp									
Granular	**			*		**	***		
Sticky				*					
Dry	*				*				
Slippery	***	*		*			***	*	
Wild			*						*
Mild								*	
Exciting				*		*			*
Staid								*	
Cool									
Kawaii				*					

***: $p < 0.001$, **: $p < 0.01$, *: $p < 0.05$

Table 3.6 ANOVA for bead-diameter, resin material, and gender of 20s

Adjective	Visual			Tactile			Both		
	Diameter	Material	Gender	Diameter	Material	Gender	Diameter	Material	Gender
Smooth	***	***		***			***		
Damp	***		**	**			***		
Granular	***	***		***			***		
Sticky									
Dry	***	**		***		**	***		**
Slippery	***			***			***		***
Wild	***	***		***			***	*	
Mild	***	***		***		***	***	**	
Exciting	***	***		***			***	***	**
Staid	***	***	**	***			***	*	**
Cool									
Kawaii	***		***	***			***		**

***: $p < 0.001$, **: $p < 0.01$, *: $p < 0.05$

As for kawaii, bead hardness had a significant main effect in the tactile presentation ($p < 0.05$). Softer beads have higher kawaii scores. Table 3.6 shows the following:

- Bead diameter has a significant main effect for kawaii for all presentation ways ($p < 0.001$). The smaller bead has the higher kawaii score.
- Gender has a significant main effect for kawaii for the visual and both presentation ways ($p < 0.001$, $p < 0.01$). The female scores are higher than those of the males.

The intersection results were omitted from this description to avoid the argument of the interpretation.

Table 3.7 shows the ANOVA results for each adjective with bead hardness, the resin material of the beads, and gender as factors for the 40s/50s. Table 3.8 shows the ANOVA results for each adjective with bead diameter, the resin material of the beads, and gender as factors for the forties/fifties.

As for kawaii, bead hardness has a significant main effect in visual presentation ($p < 0.01$). Softer beads have higher kawaii scores. The resin material of the beads also has a significant main effect in visual presentation ($p < 0.01$). Table 3.8 shows the following:

- Bead diameter has a significant main effect for kawaii for all the presentation ways ($p < 0.001$). Smaller beads have higher kawaii scores.
- Gender has a significant main effect for kawaii for the visual and both presentation ways ($p < 0.001$). The scores by females were higher than those of the males.

These results are almost the same as the results for the 20s. The intersection results were also omitted from this description.

Table 3.9 shows the results of the difference test between the groups of 20s and 40s/50s. Sticky, dry, and staid show significant differences by ages for all presentation ways. Slippery and mild have no significant differences.

We estimated the coefficients based on the multiple regression model shown in Eq. 3.1 :

$$y = a + b \log x_1 + c x_2 + d x_3 + e x_4 + f x_5 + g x_6 \qquad (3.1)$$

where

y Expected score for each adjective pair
x_1 Bead diameter (μm)
x_2 Dummy variable for bead material VI
x_3 Dummy variable for bead material FP
x_4 Dummy variable for bead hardness H
x_5 Dummy variable for bead hardness SH
x_6 Dummy variable for gender (M:0, F:1)

The following are the estimated values for kawaii for the 20s (Eq. 3.2) and the 40s/50s (Eq. 3.3):

$$a = 4.45, \ b = -1.46, \ c = 0.31, \ d = 0.13, \ e = -0.21,$$
$$f = 0.33, \ g = 0.30 \qquad (3.2)$$

Table 3.7 ANOVA for bead-hardness, resin material, and gender of 40s/50s

Adjective	Visual			Tactile			Both		
	Hardness	Material	Gender	Hardness	Material	Gender	Hardness	Material	Gender
Smooth	*	***	*	**			*		
Damp		*							
Granular		**		**			***		*
Sticky									*
Dry								*	
Slippery	***	***		**			**		**
Wild			*						
Mild		**			**				
Exciting		***							
Staid	**								
Cool			*				*		
Kawaii	**	**							

***: $p < 0.001$, **: $p < 0.01$, *: $p < 0.05$

Table 3.8 ANOVA for bead-diameter, resin material, and gender of 40s/50s

Adjective	Visual			Tactile			Both		
	Diameter	Material	Gender	Diameter	Material	Gender	Diameter	Material	Gender
Smooth	***		***	***	*	***	***		*
Damp	***						***		
Granular	***	*		***	*		***	*	
Sticky									
Dry	***	**		***			***		
Slippery	***		**	***		***	***		***
Wild	***	***		***			***	*	
Mild	***	**	***	***			***		
Exciting	***	***		***			***	***	
Staid	***	*		***		*	***		
Cool	***	*				***		*	***
Kawaii	***		**	**			***		**

***: $p < 0.001$, **: $p < 0.01$, *: $p < 0.05$

Table 3.9 Results of difference test between generations

Adjective	Visual	Tactile	Both
Smooth			*
Damp			*
Granular		**	**
Sticky		***	***
Dry	**	***	*
Slippery		**	
Wild		*	
Mild			
Exciting			*
Staid	***	*	**
Cool		*	***
Kawaii	**	*	

***: $p < 0.001$, **: $p < 0.01$, *: $p < 0.05$

$$a = 5.05, \ b = -1.83, \ c = 0.28, \ d = 0.23, \ e = 0.06,$$
$$f = 0.21, \ g = -0.06 \tag{3.3}$$

From these results, we obtained the following:

- The kawaii effects for resin material and bead hardness have the same order.
- The kawaii effects for gender only have the same order as the above factors in the 20s.
- The kawaii effect of the log of the bead diameter is larger than the above factors.

3.2.5 Summary

Experiments were performed for the affective evaluation for both visual and tactile material perception of bead-coated resin surfaces in which the resin materials, diameters, and hardness of beads were expanded systematically.

Employing 20s and middle-aged participants, we experimented on the affective evaluation of the samples of bead-coated resin surfaces to clarify the differences among different physical attributes of resin surfaces. The relation between the physical attributes of beads and the affective evaluation results was clarified from ANOVA and multiple regression analysis results.

3.3 Preparation of Adjective Pairs to Clarify the Color Effect

3.3.1 Background

We found that affective evaluations are changed by physical attributes in the previous section. However, few studies have focused on color in surface-material perception research using visual and tactile sensations. We assumed that the hues in the three elements of color change the appearance of materials and affect their affective evaluation.

This study systematically researched the affective evaluations of surface-material perception using visual and tactile sensations. We chose the number of adjective pairs from previous research to clarify the relations among surface-material perceptions with different colors and adjectives pairs. This section introduces our experimental results to reduce the excessive number of adjective pairs for affective evaluations.

3.3.2 Experimental Method

3.3.2.1 Experimental Setup

We prepared four images of colored cylinders (Fig. 3.3). The white cylinder is the standard, and the others are comparison targets. We employed three comparison targets. As indicated in Fig. 3.4, we presented images to compare a standard cylinder with cylinders of the comparison target. The background is an achromatic color with a brightness of 8 based on the Munsell color system. Adjective pairs in Japanese were displayed between the standard and the comparison targets.

3.3.2.2 Experimental Conditions

The experimental conditions are shown in Fig. 3.5 and Table 3.10.

Table 3.10 Experimental conditions

Property	Value
Display size	22 inches wide
LCD display angle	30°
Distance between eye and LCD display	50–60 cm
Room condition	Dark

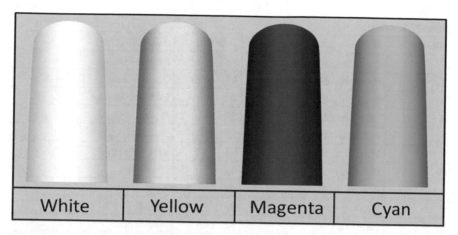

Fig. 3.3   Image of colored cylinders

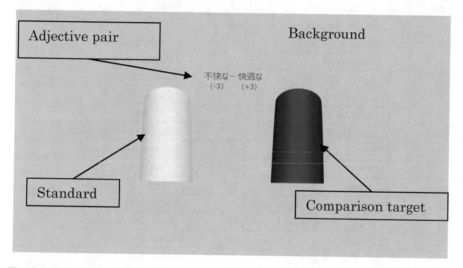

Fig. 3.4   Presented image

3.3.2.3   Adjective Pairs

We chose 61 adjective pairs from previous research on visual and tactile sensations [1, 4–7]. Table 3.11 lists the adjective pairs used for our affective evaluation experiment for which we used a 7-point Likert scale.

3.3.2.4   Experimental Procedure

Our experimental procedures were as follows:

Table 3.11 Adjective pairs

Number	Adjective pairs	Number	Adjective pairs
1	Comfortable–Uncomfortable	32	Active–Passive
2	Favorite–Hated	33	Quiet–Noisy
3	Interesting–Boring	34	Cheerful–Dismal
4	Chic–Ordinary	35	Heavy–Light
5	Artificial–Natural	36	Beautiful–Ugly
6	Masculine–Feminine	37	Lively–Sluggish
7	Hot–Cold	38	Good–Bad
8	Simple–Gaudy	39	Kind–Unkind
9	Exciting–Boring	40	Intense–Calm
10	Cute–Ugly	41	Fun–Painful
11	Dry–Sticky	42	Flashy–Sober
12	Smooth–Rough	43	Sharp–Dull
13	Relaxed–Anxious	44	Stabile–Unbalanced
14	Smooth–Prickly	45	Dynamic–Static
15	Mild–Severe	46	Rational–Irrational
16	Wild–Tame	47	Round–Square
17	Healthy–Sick	48	Clear–Obscure
18	Secure–Insecure	49	Full–Empty
19	Damp–Desiccated	50	Healing–Non-healing
20	Youthful–Aged	51	Pleasant–Unpleasant
21	Juicy–Dry	52	Gentle–Scary
22	Unique–Common	53	Clean–Unclean
23	Conspicuous–Inconspicuous	54	Refreshing–Unrefreshing
24	Cool–Old-fashioned	55	Delicious–Terrible
25	Clean–Dirty	56	Pure–Impure
26	Delicious–Disgusting	57	Powerful–Impotent
27	Resilient–Slack	58	Expensive–Cheap
28	Light–Dark	59	Fashionable–Somber
29	Moist–Dry	60	Fine in the touch–Coarse in the touch
30	Soft–Hard	61	Rare–Common
31	Warm–Cool		

Fig. 3.5 Experimental setup

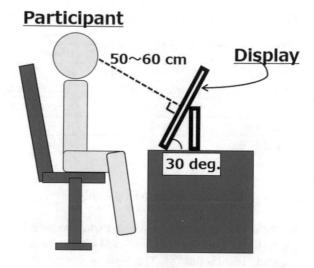

1. Participants watched background images of the display for 15 s.
2. They were presented images of the standard and comparison cylinders. The standard cylinder was always on the left of the comparison cylinder.
3. They orally evaluated their comparative impressions of the two presented cylinders for adjective pairs in Japanese on a 7-point Likert scale from −3 (strongly agree) to +3 (strongly agree).
4. They were presented images of other comparison cylinder and repeated the affective evaluations. They evaluated 61 × 3 times in total. We randomly changed the order of the comparison cylinders to cancel order effects.

3.3.3 Experimental Results

We performed our experiments with six Japanese males with normal color vision in their 20s.

3.3.3.1 Correlation Analysis

Our correction analysis results show significant correlations between all adjective pairs excluding wild pair and artificial pair. Correlations between "Warm–Cool" and some adjective pairs are shown in Table 3.12, and correlations among "Wild–Tame," "Artificial–Natural," and some adjective pairs are shown in Table 3.13.

Warm pair and hot pair have a significant positive strong correlation. Warm pair and masculine pair have a significant negative correlation. On the other hand, wild pair and artificial pair have a nonsignificant correlation.

Table 3.12 Correlations between "Warm–Cool" and some adjective pairs

	Hot–Cold	Intense–Calm	Masculine–Feminine
Warm–Cool	0.974**	0.899**	−0.645**

Table 3.13 Correlations between "Wild–Tame," "Artificial–Natural," and some adjective pairs

	Hot–Cold	Intense–Calm	Masculine–Feminine
Wild–Tame	−0.148	0.238	−0.174
Artificial–Natural	−0.148	−0.079	0.000

3.3.3.2 Hierarchical Clustering Analyses

We performed hierarchical clustering analyses with Ward's method, a complete linkage method, and a centroid method of distance metrics to confirm the robustness of the analytical results [8]. The adjectives were divided into 13 groups by Ward's method and the centroid method and 11 groups by the complete linkage method.

Based on these results of hierarchical clustering analyses, we classified the adjectives pairs into groups based on the following rules:

– The adjective pairs were categorized into identical groups regardless of metrics.
– Other adjective pairs.

Table 3.14 shows the final grouping with 11 groups and the others. From the results of these analyses, we selected adjective pairs based on one of the following conditions:

– Strongly positively correlated with many other adjective pairs in each group.
– Weakly correlated with most other adjective pairs.

Finally, we selected the 21 adjective pairs shown in Table 3.14.

3.3.4 Summary

We continue to systematically research the affective evaluations of surface-material perception using visual and tactile sensations. We experimentally reduced the number of adjective pairs for affective evaluations for subsequent experiments to clarify color effects and selected 21 adjective pairs from our experiment results. Future work will select colors for affective evaluations and experiment with them on the surface-material perception of colored bead-coated resin using visual and tactile sensations.

Table 3.14 Groups of adjective pairs (boldface shows finally selected adjective pairs)

Group1	Group2	Group3
Clean – Unclean,	Healthy – Sick,	Kind – Unkind,
Pure – Impure,	Refreshing – Unrefreshing,	Round – Square,
Comfortable – Uncomfortable,	**Relaxed – Anxious,**	Stabile – Unbalanced,
Simple – Gaudy,	Quiet – Noisy,	Secure – Insecure,
Masculine – Feminine,	Rational – Irrational	**Gentle – Scary,**
Resilient – Slack,		**Pleasant – Unpleasant,**
Beautiful – Ugly,		Mild – Severe
Delicious – Disgusting,		
Fine in the touch – Coarse in the touch		
Group4	Group5	Group6
Favorite – Hated,	Smooth – Prickly,	Full – Empty,
Healing - Non-healing,	Soft – Hard,	**Fashionable – Somber,**
Cool · Old-fashioned,	Fun – Painful,	Interesting – Boring,
Youthful – Aged,	Clean – Dirty,	Chic – Ordinary
	Good – Bad,	
	Smooth – Rough	
Group7	Group8	Group9
Damp – Desiccated,	**Light - Dark**	Hot – Cold,
Juicy – Withered,		**Warm – Cool**
Moist · Dry		
Group10	Group11	The others
Cheerful – Dismal,	**Conspicuous –**	**Sharp – Dull,**
Lively – Sluggish,	**Inconspicuous,**	**Cute – Ugly,**
Exciting – Boring,	Flashy – Sober,	**Expensive – Cheap,**
Active – Passive,	**Unique – Common,**	**Delicious – Terrible,**
Intense – Calm,	Clear – Obscure,	**Wild – Tame,**
Powerful – Impotent,	Rare – Common	**Artificial – Natural,**
Dynamic – Static		**Heavy – Light**

3.4 Combined Effect of Tactile Sensation and Hue

3.4.1 Background

In experiments that clarified the combined effect of tactile sensation and hues on the affective evaluations of surface-material perception with bead-coated resin samples and 3D models, we performed two preliminary experiments: the selection of appropriate adjective pairs for affective evaluation of the effect of hues, and the selection of hue candidates for our primary experiment [4]. In our first experiment (Sect. 3.3), we selected 21 adjective pairs, such as masculine–feminine and relaxed–anxious from the results of both correlation and hierarchical clustering analyses.

In our second experiment [4], we prepared 13 images of colored cylinders. The white cylinder was the standard. The colors of comparison targets were red, yellowred, yellow-green, green, blue-green, blue, blue-purple, purple, red-purple, gray, black, and brown. We presented two cylinder images to compare the standard and target cylinders on a PC display. The background was an achromatic color with a brightness of 8 based on the Munsell color system, which was the same as the first experiment. We displayed Japanese adjective pairs between the standard and comparison targets. We chose 12 pairs from previous research on visual sensations, beautiful–ugly, pleasant–unpleasant, and good–bad, for example. Participants watched a pair of cylinder images and a pair of adjectives, and orally evaluated the impression of comparison target on a 7-point Likert scale. They repeated this evaluation for all of the adjective pairs and all of the comparison targets, and made 144 evaluations (12×12). We randomly changed the order of the comparison cylinders to cancel any order effects, and performed our experiments with 12 Japanese males with normal color vision in their 20s. From the results of both hierarchical clustering and decision tree analyses, we selected seven colors, such as yellow-red and yellow-green.

This section describes our primary experiment based on our two preliminary experiments mentioned above.

3.4.2 Experimental Method

3.4.2.1 Samples and Models

Our experimental setup is shown in Fig. 3.6. For the affective evaluations, the participants touched the bead-coated resin surfaces of the samples in the real environment while watching 3D models in the virtual environment.

The samples are column-shaped objects with a resin side surface (Fig. 3.7). The standard sample's resin had no bead coating, but the resin of evaluation target samples was bead-coated. We employed three bead diameters: 10 μm (FP-2), 22 μm (FP-4), and 93 μm (FP-7) micrometer. Each sample pair consists of standard and target samples that are fixed by magnets (Fig. 3.8) to be touched by the participants.

The hue of the standard 3D model is white, and the hues of the evaluation target models are shown in Table 3.15, based on the result of our second preliminary experiment. Each 3D model pair consists of standard and target models displayed on gray background screen (Fig. 3.9), shown stereoscopically through a head-mounted display (Oculus rift dk2, HMD hereafter). We also showed a 3D model of the participants' right hands so that they can feel that their own right hand is touching the column's side surface. The following is the right-hand model's motion:

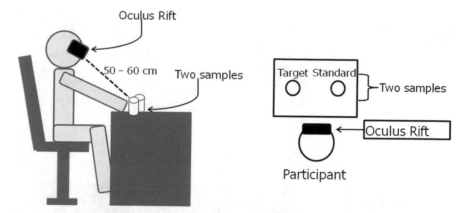

Fig. 3.6 Experimental setup

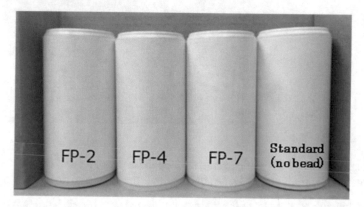

Fig. 3.7 Samples

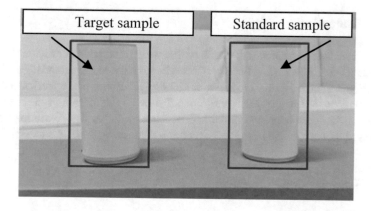

Fig. 3.8 Display of samples in real environment

Table 3.15 RGB values for hues

Hue	R	G	B
Yellow-red	243	121	9
Yellow-green	128	197	34
Blue-purple	48	48	146
Black	35	35	35
Red-purple	163	13	76
Gray	160	160	160
White	255	255	255

Fig. 3.9 Display of 3D models in the virtual environment

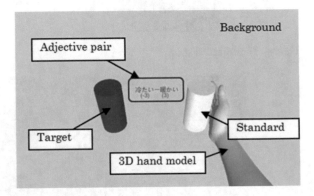

1. Stroking the standard model's side surface three times up and down.
2. Stroking the target model's side surface three times up and down.

After this, the participants give their evaluation scores for each adjective pair.

3.4.2.2 Evaluation

We used the 21 adjective pairs in Table 3.16 that were selected in our first preliminary experiment (Sect. 3.3). Each adjective pair is shown between the standard and target models (Fig. 3.3) in the virtual environment. The evaluation is performed on a 7-point Likert scale from −3 to +3. Each participant followed the motion of the right-hand model by his/her own hand and orally scored each adjective pair. Participants were also allowed to answer "I can't judge" in addition to scores from −3 to +3.

3.4.2.3 Experimental Procedure

The following is the experimental procedure:

1. Participants wore HMDs and watched a gray (brightness 8) background for 15 s.
2. They were shown both standard and target models and an adjective pair.

Table 3.16 Adjective pairs

Number	Adjective pair	Number	Adjective pair
1	Masculine–Feminine	12	Warm–Cool
2	Beautiful–Ugly	13	Active–Passive
3	Heavy–Light	14	Unique–Common
4	Gentle–Scary	15	Conspicuous–Inconspicuous
5	Pleasant–Unpleasant	16	Sharp–Dull
6	Healing–Non-healing	17	Cute–Not cute
7	Good–Bad	18	Expensive–Cheap
8	Fashionable–Sober	19	Delicious–Terrible
9	Moist–Dry	20	Wild–Tame
10	Bright–Dark	21	Artificial–Natural
11	Stable–Unbalanced		

3. They moved their own right hands that followed the 3D hand model's motion to touch the surfaces of the two samples one by one.
4. They scored the target by comparing it with the standard for the adjective pair.
5. They repeated processes 2–4 for each adjective pair. The number of repetition was 21 because the number of the adjective pairs was 21.
6. After changing the target model's hue, processes 2–5 were repeated seven times because the number of hues was 7.
7. After changing the resin of the target sample's material, processes 2–6 were repeated three times because the number of materials was 3.

The presented order of the hues and resins was set randomly.

3.4.3 Experimental Results

Twelve Japanese females in their 20s and twelve Japanese males in their 20s volunteered for our primary experiment. The experimental results were analyzed by ANOVA and regression analysis.

3.4.3.1 ANOVA Results

The evaluation results for each adjective pair were analyzed by a 3-factor ANOVA. The between-subject factor was gender, and within-subject factors were hue and bead diameter. The "masculine–feminine" result shows that hue and bead diameter have a significant main effect at a 1% level. Table 3.17 summarizes the ANOVA results. We obtained the following observations:

Table 3.17 ANOVA results

Adjective pair	Diameter	Diameter *Hue	Diameter *Gender	Hue	Hue*Gender	Gender
Masculine	**			**		
Beautiful	**		**	**		
Heavy	**			**		
Gentle	**		*	**		
Pleasant	**		*	**		*
Healing	**	*		**		
Good	**			**		
Fashionable	**		**	**		
Moist	**			**		
Bright	**			**		
Stable	**			**		
Warm				**		**
Active	*			**		
Unique	**			**		
Conspicuous	*			**		
Sharp			*	**		
Cute	**	*		**		
Expensive	**			**		
Delicious	**			**		
Wild	**			**		*
Artificial	*			**		

Significant level *5% **1%

- Hue has significant main effect for all adjective pairs at a 1% level.
- Bead diameter has a significant main effect for all adjective pairs except "warm—cool" and "sharp–dull" at a 5% level.
- Gender has a significant main effect for "pleasant–unpleasant," "warm–cool," and "wild–tame" at a 5% level.

These results suggest that hue and bead diameter have a stronger effect on affective evaluation than gender.

Then, we focused on the relation between hue and bead diameter, performed a multiple comparison, and obtained the following observations:

- The "masculine–feminine" pair has a lower score (meaning that it is more masculine) for red-purple than other hues at a 5% level.
- The "beautiful–ugly" pair, "pleasant–unpleasant" pair, and some other pairs have lower scores for larger bead diameter than the other diameters at a 5% level.

3.4.3.2   Regression Analysis Results

We created the following regression formula by employing a common logarithm of bead diameter, hue, and gender as variables.

$$y = a + b\log x_1 + cx_2 + dx_3 + ex_4 + fx_5 + gx_6 + hx_7 + ix_8 \qquad (3.4)$$

where

y Expected score for each adjective pair,
x_1 Bead diameter (μm),
x_2 Dummy variable for yellow-red (1: yellow-red, 0:otherwise),
x_3 Dummy variable for yellow-green (1: yellow-green, 0:otherwise),
x_4 Dummy variable for blue-purple (1: blue-purple, 0:otherwise),
x_5 Dummy variable for red-purple (1: red-purple, 0:otherwise),
x_6 Dummy variable for gray (1: gray, 0:otherwise),
x_7 Dummy variable for black (1: black, 0:otherwise),
x_8 Dummy variable for female (1: female, 0:otherwise).

We estimated the partial regression coefficients for each adjective pair. The following are the coefficients for the "masculine–feminine" pair:

$$a = -3.31, \; b = 0.46, \; c = -0.1, \; d = 0.1, \; e = 0.27, \; f = -0.26, \; g = 2.1,$$
$$h = 0.2, \; i = -0.05.$$
$$(3.5)$$

This result shows that masculine scores increased when the participant is male, the hue is blue-purple, and bead diameter is large. Table 3.18 shows the partial regression coefficients for all adjective pairs. We obtained the following observations:

- The larger the bead diameter is, the higher are "masculine," "heavy," "aggressive," "unique," "conspicuous," "wild," and "artificial" scores.
- Females gave higher scores for "beautiful," "calm," "gentle," "relaxed," "heavy," "warm," and "expensive."

Hue, bead diameter, and gender influence the scores of adjective pairs.

3.4.4   Discussion and Summary

We clarified the combined effect of tactile sensation and hues on the affective evaluations of surface-material perception by performing an affective evaluation experiment with 21 adjective pairs. We used 3D models with hues on their side surfaces in virtual environment and cylindrical samples with bead-coated resin on their side surfaces

Table 3.18 Multiple regression analysis results

Adjective pair	Constant	Diameter	YR	YG	BP	RP	Gray	Black	Female	
Y	a	b	c	d	e	f	g	h	i	R^2
Masculine	−3.31**	0.46**	−0.10**	0.10**	0.27*	−0.26**	0.21**	0.20**	−0.05	0.46**
Beautiful	3.13**	−0.54**	0.49	0.01	0.03	0.03	0.18**	−0.15**	0.09*	0.35**
Heavy	−2.38**	0.31**	−0.08	−0.09	0.18**	0.11*	0.24**	0.34**	0.04	0.28**
Gentle	3.40**	−0.48**	0.15**	0.05	−0.15**	−0.05	−0.11*	−0.3**	0.06	0.37**
Pleasant	2.55**	−0.46**	0.20**	0.1*	−0.05	−0.01	−0.1*	−0.13**	−0.11**	0.31**
Healing	4.08**	−0.53**	0.00	0.05	−0.08	−0.19	−0.13**	−0.25**	0.03	0.36
Good	3.33**	−0.50**	0.06	0.06	−0.06	−0.12*	−0.15**	−0.22**	0.05	0.32
Fashionable	1.50**	−0.25**	0.11*	0.02	−0.01	0.17**	−0.28**	−0.09	0.09*	0.20**
Moist	1.04**	−0.32**	0.12*	0.22**	0.19**	0.15**	−0.15**	−0.11*	−0.06	0.24**
Bright	0.67**	−0.11**	0.24**	0.12**	−0.26**	0.03	−0.30**	−0.48**	−0.07*	0.47**
Stable	3.81**	−0.52**	−0.09	−0.20	0.05	−0.18**	0.02	−0.07	0.04	0.3**
Warm	0.32	−0.04	0.30**	0.04	−0.32**	0.17	−0.13**	−0.1*	0.15**	0.30**

(continued)

Table 3.18 (continued)

Adjective pair	Constant	Diameter	YR	YG	BP	RP	Gray	Black	Female	
Active	-0.70^{**}	0.09^{*}	0.28^{**}	0.1^{*}	-0.18^{**}	0.3^{**}	-0.24^{**}	-0.16^{**}	-0.03	0.34^{**}
Unique	-1.49^{**}	0.29^{**}	0.19^{**}	0.05	0.03	0.19^{**}	-0.23^{**}	0.02	-0.04	0.21^{**}
Conspicuous	-1.34^{**}	0.17^{**}	0.27^{**}	0.16^{**}	-0.02	0.29^{**}	-0.2^{**}	0.00	-0.06	0.24^{**}
Sharp	-1.07^{**}	0.01	0.08	-0.00	0.06	0.17	-0.17^{**}	-0.06	-0.05	0.04^{**}
Cute	2.60^{**}	-0.41^{**}	0.17^{**}	0.05	-0.16^{**}	0.05	-0.21^{**}	-0.30^{**}	-0.09^{*}	0.37^{**}
Expensive	2.44^{**}	-0.37^{**}	-0.13^{*}	-0.16^{**}	0.01	-0.21	-0.19^{**}	-0.01	0.11^{**}	0.18^{**}
Delicious	2.05^{**}	-0.36^{**}	0.24^{**}	0.07	-0.16^{*}	0.03	-0.22^{**}	-0.26^{**}	-0.05	0.35^{**}
Wild	-1.93^{**}	0.42^{**}	-0.09	-0.03	-0.05	-0.06	0.12^{*}	0.17^{**}	-0.16^{**}	0.26^{**}
Artificial	-1.21^{**}	0.12^{**}	0.04	-0.11	0.10	0.26^{**}	0.13^{*}	0.17^{**}	-0.04	0.10^{**}

Significant level $^{*}5\%$ $^{**}1\%$

in a real environment. From analysis of the experimental results by ANOVA and multiple regression analysis, we obtained the following conclusions:

– Hue affects evaluations of all adjective pairs in Table 3.17.
– Bead diameter affects the evaluations for most adjective pairs, except "warm—cool," and "sharp–dull."
– Gender affects the evaluations for such adjective pairs as "pleasant–unpleasant," "warm–cool," and "wild–tame."

We also obtained multiple regression formulas for each adjective pair. Future work will consider the differences between generations.

The results can be applied to emphasizing the impressions of contents by their packages such as tea cans and cosmetic bottles.

3.5 Conclusion

We performed affective evaluation experiments on both visual and tactile material perception of bead-coated resin surfaces in which the materials, the diameters, and the hardness of beads were expanded systematically.

After two preliminary experiments on the similarity of the bead-coated resin surfaces to reduce number of material samples and on the similarity of the evaluation items to reduce their number, we performed an affective evaluation experiment on selected samples with students and middle-aged participants to clarify the differences among the different physical attributes of resin surfaces and focused on kawaii. The relation between the physical attributes of beads and the affective evaluation results were clarified from ANOVA and multiple regression analysis results.

Then, we added hues, which is one of the three elements of color, and performed two preliminary experiments to select appropriate adjective pairs for affective evaluation of the effect of hues and to select the hue candidates. With 21 adjective pairs, and 7 hues, we clarified the combined effect of the tactile sensation and the hues on the affective evaluations using combinations of actual bead-coated resin and 3D models. Our analysis results suggest that hue, bead diameter, and gender are important for affective evaluation. We also obtained multiple regression formulas for each adjective pair.

The results can be applied to emphasizing the impressions of contents by their packages such as tea cans and cosmetic bottles.

Acknowledgements We would like to express our sincere thanks to all participants in our experiments.

References

1. Ohkura, M., et al., (2015). Affective evaluation for material perception of bead-coated resin surfaces using visual and tactile sensations-Focusing on kawaii. In *Proceedings of ISASE2015*.
2. Morishita, W., et al. (2016). Affective evaluation for material perception of bead-coated resin surfaces using visual and tactile sensations: Preparation of adjective pairs to clarify the color effect. In *Proceedings of APD2016* (pp. 251–258).
3. Ohkura, M., et al. (2017). Analysis of affective evaluation for material perception of resin surfaces: Combined effect of tactile sensation and hue. In *Proceedings of APD2017* (pp. 190–200).
4. Morishita, W., et al. (2016). Analysis of affective evaluation for material perception of resin surfaces–classification of hues for affective evaluations. In *Proceedings of the 11th Spring meeting of Japan Society of Kansei Engineering* (G13-1). (in Japanese).
5. Inoue, M., & Kobayashi, T. (1985). The research domain and scale construction of adjective-pairs in a semantic differential method in Japan. *The Japanese Journal of Educational Psychology, 33*(3), 253–260. (in Japanese).
6. Soeta, Y., Kitamoto, T., Hasegawa, H. (2015) IEICE transactions on fundamentals of electronics. *Communications and Computer Sciences, J98*-A(6), 436–445. (in Japanese).
7. Suzuki, M., & Gyoba, J. (2002). Contrastive analysis of sensory: Relevance of factors affecting aesthetic impressions. *Technical Report of IEICE HIP, 101*(698), 31–38. (in Japanese).
8. Joe H. Ward, Jr. (1963). Hierarchical grouping to optimize an objective function. *Journal of the American Statistical Association, 58*(301), 236–244.

Chapter 4
Kawaii-ness in Motion

Shohei Sugano and Ken Tomiyama

Abstract This chapter proposes a concept of KANSEI mathematical model and reports our attempts to find out physical attributes that determine Kawaii-ness in motion. A Japanese word "Kawaii" is one of the representative concepts of Japan-original KANSEI (emotions and sensibility) and describes favorable characters such as pretty, adorable, fairy-like, and cute. Although there are many attempts to clarify Kawaii-ness in static quantities such as shapes and colors, there is none in motions. We have been developing Virtual KANSEI for robots that enable robots to communicate with humans not only at a physical level but also at a metaphysical level. So far, we have succeeded in developing virtual emotion but not sensibility. This chapter extends our research on virtual emotion to virtual sensibility. First, we proposed the concept of KANSEI mathematical model. Then, we attempted to find influential factors that determine Kawaii-ness of motion. We chose Roomba to study Kawaii-ness of motion because its neutral shape (round) and color (white) are expected to give little influence on Kawaii-ness of motion. We first investigated impressions of motions of Roomba and classified them into 10 types. Then, we used a questionnaire consisting of 20 pairing adjectives to evaluate those motions. We conducted a factor analysis using the top three motions that were found to be Kawaii by the participants. We found a factor that is common to all three motions that could be named infantile. The result of the factor analysis also suggests several physical quantities that control Kawaii-ness of motion.

Keywords Kawaii · Robot motion · Virtual KANSEI

S. Sugano (✉)
aba Inc., Funabashi, Japan
e-mail: shohei.sugano@gmail.com

K. Tomiyama
Chiba Institute of Technology, Narashino, Japan

4.1 Introduction

The objective of our study is to propose a KANSEI mathematical model that describes Kawaii-ness of motion. It is a part of ongoing research of Virtual KANSEI for robots [1–4]. Here, we present our initial attempts to identify physical quantities that have influence on Kawaii-ness of motion. Many researchers have studied Kawaii-ness. For example, there are studies on Kawaii-ness in terms of color, shape, size, and texture [5–7]. Some studies pay attention to biological signals or look at implications from a behavioral science point of view [8, 9]. The readers are referred to other articles of this book for the results in those areas. However, there is little research on Kawaii-ness in motion and that is the subject of this study.

We all find Kawaii-ness in moving objects, but it is unclear what kind of physical attributes makes a motion Kawaii. Two standing parameters of motion are speed and acceleration. However, even those have six dimensions each in a 3D space, namely, for 3D translational and rotational motions. We may need to consider the third derivative of position called the jerk in six dimensions. We also may need to consider the influence of curvature and complexity of robot trajectories. Thus, the number of combinations of those parameters could be prohibitable.

We decided to limit the motion in a 2D plane for this study. Moreover, since the motions need to be duplicated for multiple experiments with the same condition, we adopted Roomba by iRobot as our robot [10]. The motion of Roomba is known to be constructed from a few simple component motions as we explain later. It is also considered that the round shape of Roomba with minimal texture and uniform color can be assumed not particularly Kawaii and would have little influence on Kawaii-ness of motion. This point was clarified in our experiment.

4.2 KANSEI and Emotion

Although emotion is considered as a part of KANSEI (Fig. 4.1), characteristics of emotion and KANSEI are quite different.

Fig. 4.1 Relationship between KANSEI (sensibility) and KANJO (emotion)

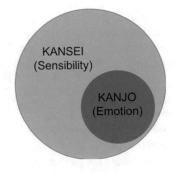

Fig. 4.2 The emotion estimation by facial expression is possible

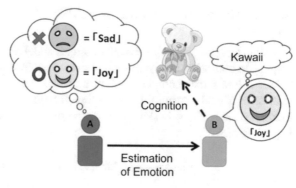

Fig. 4.3 The KANSEI estimation by facial expression is impossible

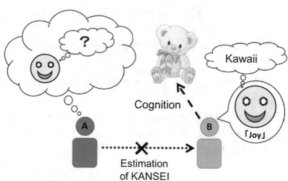

Emotion of a person can be detected by observing physical indications of a person such as voice sounds, facial expressions, and body motions. However, KANSEI cannot be observed unless it is declared by the person. This can be understood easily by looking at an example where person A is watching person B who just received a teddy bear from person C (Figs. 4.2 and 4.3). Person A can find that person B is happy by looking at, for example, the angle of the edges of the mouth. But person A cannot conclude that person B finds the bear Kawaii unless person B declares so.

Therefore, robots can detect the emotion state of a person by obtaining easily available physical signals such as facial expression, body movements, and voice sounds. It is also possible to express a robot emotion by simple physical attributes such as intentional movement of parts of the robot body. That was the reason we started from emotion studies. We have proposed the concept of Virtual KANSEI and three module structures of the Virtual Emotion and developed three modules: an emotion detection module to detect the emotion state of the partner, an emotion generation module to generate the virtual emotion of the robot, and an emotive motion modulation module for modulating a robot motion for a given task.

In 2010, we started studying KANSEI itself and focused on Kawaii as our target KANSEI because Kawaii is a Japanese original KANSEI concept that represents cute and other favorable characteristics and enjoys worldwide recognition and popularity.

4.3 Mathematical Model of KANSEI

Our main target is to develop a mathematical model of KANSEI. KANSEI is how we respond to both stimulations from the external environment and internal unconscious states. It is a rather complex system that may require a nonlinear multi-variable relationship involving multitude of inputs (factors) that are representing external and internal stimuli (Eq. 4.1) (Fig. 4.4). For example, Kawaii-ness can be expressed by Eq. 4.2 (Fig. 4.5).

$$Kansei = f(factor_1, factor_2, factor_3, \ldots) \tag{4.1}$$

$$Kawaii = g(shape, color, texture, motion, \ldots) \tag{4.2}$$

Since this study is focused on Kawaii-ness induced by motion, we may express

$$Kawaii_{motion} = \frac{\partial g}{\partial motion}(attribute_1, attribute_2, attribute_3, \ldots) \tag{4.3}$$

where attributes in this equation represent physical quantities that specify characteristics of motions. This equation can be further rewritten by expanding in terms of motion attributes as

$$Kawaii_{motion} = \sum h_i(attribute_i) \tag{4.4}$$

where h_i are partial derivatives of $Kawaii_{motion}$ with respect to $attribute_i$.

Thus, our task is to find this equation. It is a fundamental fact of the partial derivative that when differentiating with respect to a variable, all other variables must be kept constant. This is a severe constraint for our study because motions are specified by combinations of many quantities and keeping all quantities constant except one is almost impossible. Nonetheless, we need to start from somewhere and we propose a linear model in the form of multiple regression analysis because all the

Fig. 4.4 A block diagram of KANSEI mathematical model

Fig. 4.5 A block diagram of KANSEI mathematical model for Kawaii-ness

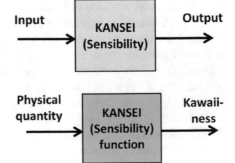

variables are free to be varied in compiling data. This choice is proposed based also on the fact that it has proven effectiveness in modeling multi-variable relationships. Thus, the proposed model is

$$Kawaii_{motion} = \sum \alpha_i \cdot attribute_i \qquad (4.5)$$

where α_i are the model parameters which are to be determined.

Several comments are due here. Since we are interested in using the model in robotics applications, it is better that the attributes in this equation are measurable quantities using simple sensors such as an accelerometer. Although this model uses physical attributes of motion directly, it may be improved if we use a logarithm of the attribute values as Weber–Fechner Law states that human perception is proportional to the logarithm of physical stimulus [11, 12]. Also commented that the model parameters can be tactically adjusted to give unique KANSEI to individual robots. This can help making our robots compatible with wide range of persons.

4.4 Adoption of Roomba

The purpose of this study is to find and analyze the characteristic features of motions that humans find Kawaii. For this purpose, we showed several video clips of Roomba motions and asked Kawaii-ness of those motions to experiment participants. We adopted Roomba for our experiment because of the following reasons:

- Rather neutral shape and color (see Fig. 4.6).
- The motion composed of three basic moves.
- Operation on a 2D plane.

Fig. 4.6 Roomba (model: 537) [13]. This robot was chosen as a testing bed

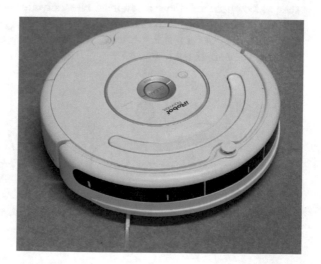

- Readily available autonomous home appliance.

As stated above, motions of Roomba are composed of three pairs of basic moves: forward and backward linear moves, right and left turns, and right and left rotations. For example, spiral motion consists of rotation and linear moves. This gives us an easy reproduction method of characteristic motions of Roomba. Also, unlike a typical robot such as AIBO, Roomba with the round shape and off-white color is not particularly Kawaii (we checked this point in our experiment) and will not affect Kawaii-ness of motion because it does not vary as Roomba moves. This is a favorable factor in studying Kawaii-ness in motion. Incidentally, many video clips of Roomba motions are uploaded on YouTube and have gained a Kawaii reputation. This also supports our adoption of Roomba for our study (Fig. 4.6).

4.5 Previous Results

We will report a series of studies we conducted and give brief explanations of them here.

4.5.1 Identification and Segmentation of Characteristic Motions of Roomba [14]

When we use the word "motion," we need to define it. Since Roomba keeps moving, we need to record its motion and divide it to identifiable segments. For this purpose, we showed a scene of Roomba in motion to nine participants and asked them to identify characteristic motions by arrow diagrams like the ones shown in Fig. 4.7 and made video clips from those diagrams. The video clips are shown to the participants and modified until they satisfactorily represent the images of motions described by the participants. As a result, we were able to identify and categorize 10 separate motions. They are listed in Fig. 4.8.

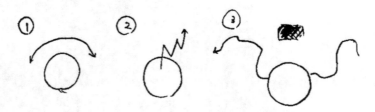

Fig. 4.7 Line drawings by a participant to indicate Kawaii motions of Roomba

Fig. 4.8 Ten segmented
motions of Roomba
classified as Kawaii

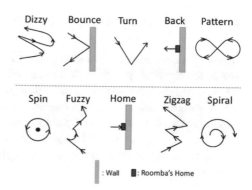

Fig. 4.9 The schematics of
an experiment to determine
the camera position and
angle for recording Roomba
motions

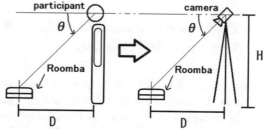

D: Distance from Roomba to the observing participant
H: Distance from the ground to the participant's eyes
θ: The view angle at which the participant watched Roomba

4.5.2 Finding a Setting of Natural Viewing of Roomba [14]

To prepare video clips for evaluation experiments of Kawaii-ness of the Roomba
motion, we measured the natural viewing angles and distances of the nine participants
from Roomba. We used the setup shown in Fig. 4.9 to determine the camera angle θ
and positions D and H. They were found to be $\theta = 45$ (deg) and D = H=155 (cm).
These values were used for recording 10 characteristic motions of Roomba for the
evaluation experiments.

4.5.3 Evaluation of 10 Recorded Roomba Motions
and Finding Kawaii Motions [14]

The prepared video clips were shown to the same nine participants as in Sects. 4.5.1
and 4.5.2 to identify Kawaii motions. The participants were asked to pick as many
motions as they found Kawaii. The motions were ordered, and the top three motions
were chosen as Kawaii motions: Bounce, Dizzy, and Spiral.

4.5.4   Preliminary Analysis of Three Roomba Motions [14]

We analyzed three Roomba motions which were found Kawaii by the participants for associated physical quantities. We extracted the following physical quantities from the Roomba motions found Kawaii: time intervals of motions, position histories, speed histories, angle (between the local axis defined by the center and the front of Roomba and the direction from the fixed viewpoint to the center of Roomba) histories, and angular speed histories. We found the speed of linear motion was closely controlled to 300 (mm/s) but the angular speed varied widely depending on the motion.

4.5.5   Evaluation of Kawaii-ness Using 20 Adjective Pairs with a 5-Point Scale [14]

The three chosen Kawaii motions of Roomba were used to further analyze their impressions. The set of 20 adjective pairs, chosen through a preliminary experiment, is listed in the appendix. We assigned, from the left, $-2, -1, 0, 1$, and 2 for the markings of the adjective pairs. We found that "Simple," "Smooth," and "Regular" were commonly chosen for Kawaii motions. We also computed the means and standard deviations of the pairs over the participants. The scores were

- *Simple*/Complex ($\sigma = 0.53$, m $= -1.5$),
- *Smooth*/Coarse ($\sigma = 0.52$, m $= -1.5$), and
- *Regular*/Irregular ($\sigma = 0.70$, m $= -1.4$).

Here, σ and m are the standard deviation and mean, respectively. A negative mean indicates the selection on the left-hand side and these are indicated by the Italic. A small standard deviation implies consistency in selection.

The participants were asked to give free comments on why the motions they chose were found Kawaii. The top three comments were "childish motion," "like a baby," and "just like my pet." However, the adjective "childish" was not particularly chosen among the adjective pairs. This anomaly may have been caused by the small number of participants, which was 18 (9 males and 9 females).

4.6   The Study of Influences of Physical Attributes on Kawaii-ness of Motion

We generated motions with controlled physical attributes and used them to investigate the influences of physical attributes on Kawaii-ness.

4.6.1 Composition of 10 Motions with Controlled Physical Attributes [15]

We prepared several basic motions with multiple speeds and accelerations using Create, a research version of Roomba, for studying influences of those physical attributes on Kawaii-ness of motion. We composed nine categories of motions, which are

- Bounce_A: Create collides obliquely against the wall and travels again after rotating 90°.
- Bounce_B: Create moves 1.5 s backward after colliding obliquely against the wall and travels again after rotating 90°.
- Spiral: Create moves along an outward spiral.
- Spin_A: Create spins around at a constant speed like a top spinning at a spot.
- Spin_B: Create spins with gradually increasing speed from 0 to 500 mm/s (LH) or gradually decreasing speed from 500 to 0 mm/s (HL).
- Straight_A: Create travels from left to right without stopping.
- Straight_B: Create travels from left to right with deceleration, stop, and acceleration phase in the middle.
- Straight_C: Create travels from left to right with a sudden stop in the middle.
- Attack: Create repeats the collision with the object at 300 mm/s.

All motions except Attack and Spin_B are composed with three speeds; 100, 300, and 500 mm/s. Thus, we composed a total of 24 types of motions.

4.6.2 Evaluation of Kawaii-ness Using 20 Pairs of Adjectives [15]

We recruited 20 new participants for this evaluation experiment. Each participant was asked to evaluate the impression of chosen motions using 20 adjective pairs. He/she was asked to choose one of four choices A, B, C, and D shown on a screen. Once a choice was made, the corresponding motion was shown to the participant. This way, the participants did not know what motions were associated with other choices. This selection was repeated 12 times so that each participant watched 12 out of 24 motions. We adopted a touch panel for the ease of selecting motions. It is noted that the association between the four letters and motions was set randomly from the remaining motions excluding those motions shown to the participant.

It was observed that although 80% of all the participants answered that Create itself was not Kawaii, 80% of them found Spiral (300 mm/s) motion Kawaii. This proves the existence of Kawaii-ness induced by motion. We found

- Rotational motions were found more Kawaii than linear motions.
- Motions with instantaneous changes in acceleration caused by collisions were evaluated more Kawaii than the motions without collisions.
- Motions that gave biological and intentional impressions were also evaluated Kawaii.

The last item was deduced from the comments of the participants to the motions they found Kawaii. This indicates that Kawaii-ness may be associated with animacy perception. Since animal motions are rich in sudden unexpected changes in physical characteristics such as the speed and the direction of motion, we built a hypothesis that changes in acceleration, namely, the jerk will affect the Kawaii-ness of motion. The second observation above supports our hypothesis. However, there are motions that are found Kawaii but do not contain a noticeable jerk, indicating the existence of other types of Kawaii-ness in motion. However, the number of participants was too small to draw definite conclusions.

4.7 Evaluation Experiment

Here, we repeated the experiments[1] in Sects. 4.5.3 and 4.5.5 with more participants and analyzed the evaluation result using the factor analysis. Ten Roomba motions shown in Fig. 4.8 were shown to a set of new participants. The number of participants this time was 52 (23 males and 29 females) and the evaluation time was 35 min. However, the number of participants with valid answers was 32 (11 males and 21 females) excluding nonvalid answers such as missing markings.

As before, 10 motion video clips were shown at random orders, and the participants were asked to answer the questionnaire only for those motions they found Kawaii. The questionnaire contained randomly ordered 20 pairs of adjectives. Three sets of questionnaires with different orders were prepared and randomly used. The adjective pairs are listed in the Appendix.

4.7.1 Results

We included several pre-experiment questions which had been administered before we showed the video clips of Roomba motions. 90% of the participants answered that they knew Roomba and 86% said they had seen a moving Roomba. However, as Fig. 4.10 shows, more than half of the participants answered no for the question "Do you think Roomba is Kawaii?" It is noted that this question is administered while showing the picture of Roomba in Fig. 4.6. This confirms our assumption that Roomba itself is not particularly Kawaii.

---

[1]This experiment was approved by the ethics committee of the Chiba Institute of Technology.

Fig. 4.10 Responses to the question "Do you think Roomba is Kawaii?"

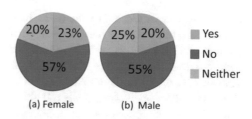

20% 23%

57%

25% 20%

55%

■ Yes
■ No
■ Neither

(a) Female   (b) Male

Unlike our previous study (Sect. 4.5.3), the top three motions chosen as Kawaii by both males and females were Bounce, Dizzy, and Fuzzy as shown below:

- 1st—Bounce (males: 11, females: 14),
- 1st—Fuzzy (males: 9, females: 16), and
- 3rd—Dizzy (males: 10, females: 13).

On the other hand, Turn, Back, Spin, and Home comprised the bottom four with less than 20% of the participants choosing it to be Kawaii. It is noted that the only difference between Bounce and Turn is the collision with the wall. This suggests the passiveness (Bounce) and activeness (Turn) may have an influence on Kawaii-ness. In other words, the existence and the lack of intention may affect Kawaii-ness.

Interestingly, although 57% of the female participants answered Roomba is not Kawaii in the pre-experiment question, 53% of them found Fuzzy motion Kawaii. Similarly, 55% of the male participants answered Roomba is not Kawaii but 55% of them found Bounce Kawaii. This indicates that there are cases where Kawaii-ness originates from the motion, not the shape of Roomba.

4.7.2 Discussion on the Markings of Adjective Pairs

We next paid attention to the deviation of choices among pairs of adjectives. The reason is because a small deviation means consistent opinions among the participants, and this implies that common characteristics of motion which contribute to Kawaii-ness of motion exist. As we have used in our previous studies, we assigned, from the left, -2, -1, 0, 1, and 2 for the markings of the adjective pairs and computed the means and standard deviations of the pairs over the participants. We found the following six pairs having small standard deviations:

- *Smooth*/Rugged ($\sigma = 0.50$, m $= -1.4$),
- Complicated/*Simple* ($\sigma = 0.62$, m $= 1.2$),
- Straight/*Curved* ($\sigma = 0.66$, m $= 1.4$),
- *Regular*/Irregular ($\sigma = 0.67$, m $= -1.2$),
- *Young*/Elder ($\sigma = 0.74$, m $= -1.3$), and
- Hasty/*Gradual* ($\sigma = 0.75$, m $= 1.3$).

Here, σ and m indicate the standard deviation and the mean, respectively. The positive/negative mean implies the preference on the right/left adjective and the chosen sides are indicated by the Italic. In addition to Simple, Smooth, and Regular, which were found in our previous study to be common choices, we found Curved, Young, and Gradual as well. The fact that there exist adjective pairs with small standard deviations implies the existence of common factors that contribute to the Kawaii-ness of motion. For example, a combination of Simple, Regular, and Young reminds undeveloped movements of babies. Smooth and Gradual can be associated with physical quantities such as acceleration and angular speed. With the smallest standard deviation, smoothness may be one of the main physical quantities humans look at in evaluating Kawaii-ness.

4.7.3 Discussion on the Results of the Factor Analysis

Next, we further analyze the top three choices, Bounce, Fuzzy, and Dizzy. The following tables are the results of the factor analysis. In those tables, r and p represent the cumulative contribution ratio and the p-value of chi-square test statistic, respectively.

The results of the factor analysis are summarized in the following three tables (Tables 4.1, 4.2, and 4.3).

Table 4.1 Bounce (factor $= 5$, $r = 60.4$; $p = 0.168 > 0.05$)

	Factor 1	Factor 2	Factor 3	Factor 4	Factor 5
First	Toddling	Short	Smooth	Grown-up	Young
Second	Stupid	Rebellious	Simple	Unknown	Clear
Third	Irregular	Straight	Idle	Natural	Unknown
Named Factor	Infantile	Agility	Harmonious	Sociability	Positiveness

Table 4.2 Fuzzy (factor $= 5$, $r = 59.1$; $p = 0.635 > 0.05$)

	Factor 1	Factor 2	Factor 3	Factor 4	Factor 5
First	Toddling	Simple	Unknown	Long	Weak
Second	Slow	Regular	Rugged	Nervous	Obedient
Third	Childish	Stubborn	Rebellious	Old	Relaxed
Named Factor	Infantile	Mechanic	Affinity	Thoughtfulness	Strength

Table 4.3 Dizzy (factor $= 5$, $r = 62.7$; $p = 0.0014 < 0.05$)

	Factor 1	Factor 2	Factor 3	Factor 4	Factor 5
First	Gradual	Toddling	Rebellious	Idle	Young
Second	Relaxed	Ambiguous	Funny	Ragged	Long
Third	Irregular	Weak	Stupid	Familiar	Gradual
Named Factor	Affinity	Infantile	Sociability	Concern	Positiveness

With the Bounce motion, the first factor was named "Infantile" from the three adjectives: Toddling, Stupid, and Irregular. The second and other factors were named "Agility," "Harmonious," "Sociability," and "Positiveness." The factors for the Fuzzy motion were "Infantile," "Mechanic," "Affinity," "Thoughtfulness," and "Strength," in this order. For the Dizzy motion, five factors may not be enough because the p-value of the chi-squared statistics of 0.0014 is smaller than 0.05 and the null hypothesis "five factors is enough" is rejected. Even though the null hypothesis is rejected, we can observe a similar tendency as other motions in those five factors, "Affinity," "Infantile," "Sociability," "Concern," and "Positiveness."

The similarity and differences in the factors of Bounce and Dizzy motions are very suggestive. The first factor of both motions is Infantile and that supports a universal recognition of Kawaii-ness in infants. The adjectives that constitute the first factor of "Infantile" of both Bounce and Fuzzy motions, namely, Toddling, Slow, and Irregular, suggest physical attributes such as speed, complexity of trajectory, and curvature. They are possible quantities for the input of the mathematical model of Kawaii.

The other factors are different but the adjectives appearing in those tables are similar, and this may indicate that the participants are looking at similar characteristics but expressing in different manner. The difference can also indicate that there are several types of Kawaii-ness. From those observations, we concluded that we do not have enough data to construct a mathematical model of Kawaii-ness. Although the number of the participants was more compared to that in our previous studies, it is not enough to draw more solid conclusions for developing a viable model.

4.8 Concluding Remarks

The objective of our study was to propose a KANSEI mathematical model that describes Kawaii-ness of motion. The target of our study was to develop a mathematical model of Kawaii-ness of motion. In the following, we present a summary of our attempts to identify physical attributes that contribute to Kawaii-ness of motion.

We chose Roomba as our experimental apparatus to identify characteristic motions because the shape and color of Roomba do not particularly induce Kawaii impression, and motions of Roomba are generated by combining few basic motions. The former is important to eliminate external influence on Kawaii-ness of motion and the latter is important for consistency in motions in repeated experiments. We found and categorized 10 motions, whose video clips were recorded in natural viewing environment. The Kawaii-ness of those motions was evaluated by a set of participants. It was found that three motions, named Bounce, Dizzy, and Spiral, were the top three choices. We further analyzed those three motions for their physical characteristics such as the speed, acceleration, jerk, and others. Then, we used Create, a research version of Roomba, to generate a total of 24 motions based on the analysis: seven types of motions with three speed settings, a rotation with ascending and descending speeds and one attacking motion. The participants were asked to evaluate randomly chosen 12 motions using a questionnaire with 20 adjective pairs.

We also repeated the experiment using 10 Roomba motions with more participants and analyzed the result using the factor analysis. The top two motions, Bounce and Dizzy, found Kawaii was the same as the result of the previous study, but the third one, Fuzzy versus Spiral, was different. The factor analysis revealed the factor "Infantile" as the top factor for the top two motions. The other factors were different but the adjectives appearing on the factor table were similar, and this implies two possibilities: individuality in evaluating Kawaii-ness among participants and existence of multiple types of Kawaii-ness in motion.

We were not able to develop a concrete mathematical model of Kawaii-ness of motion, and the main cause of this was the lack of enough data. In the experiment, we found that showing randomly chosen motions to the participants and letting the participants compare the motions in addition to answering a questionnaire not only takes time but also forces patience and efforts on the participants. This was the main cause of not being able to secure enough number of participants and sizeable nonvalid questionnaire answers. Thus, an obvious future task is to improve the way we conduct our experiments and increase the number of valid data sets. We can develop a concrete mathematical model of Kawaii-ness of motion only after we have compiled enough amount of data.

Acknowledgements The authors would like to express their appreciation for all the participants of our series of Kawaii studies for their patience and willingness for cooperation. We also would like to thank Ms. Haruna Morita who was our trusted comrade in the early stage of our Kawaii study. Special appreciation is due to Prof. Michiko Ohkura for giving us the opportunity to report our work here.

Appendix

The adjective pairs used in our questionnaire are listed below. Three different questionnaires were prepared with a randomized order of the pairs and right and left arrangements of adjectives (Table 4.4).

Table 4.4 Adjective pairs in the questionnaire

No	Adjective pairs	No	Adjective pairs
1	Obedient–Rebellious	11	Toddling–Brisk
2	Simple–Complicated	12	Childish–Mature
3	Familiar–Unknown	13	Smart–Stupid
4	Natural–Artificial	14	Relaxed–Nervous
5	Yielding–Stubborn	15	Serious–Funny
6	Long–Short	16	Smooth–Ragged
7	Quick–Slow	17	Hasty–Gradual
8	Young–Elder	18	Idle–Busy
9	Strong–Weak	19	Regular–Irregular
10	Straight–Curved	20	Clear–Ambiguous

References

1. Miyaji, Y., & Tomiyama, K. (2007). Virtual KANSEI for robots in welfare. In *Proceedings of IEEE/ICME International Conference on Complex Medical Engineering* (pp. 1323–1326).
2. Kogami, J., Miyaji, Y., & Tomiyama, K. (2009). KANSEI generator using HMM for virtual KANSEI in caretaker support robot. *KANSEI Engineering International Journal, 8*(1), 83–90.
3. Zenkyoh, M., & Tomiyama, K. (2011). Surprise generator for virtual KANSEI based on human surprise characteristics. In *Proceedings of HCI2011, The 14th International Conference on Human-Computer Interaction* (S111). Orlando, Florida, USA.
4. Miyaji, Y., & Tomiyama, K. (2003). Construction of virtual KANSEI by Petri-net with GA and method of constructing personality. In *Proceedings of ROMAN2003, The 12th IEEE International Workshop on Robot and Human Interactive Communication* 6B4(CD–ROM) (pp. 391–396). Millbrae, California, USA.
5. Ohkura, M., Konuma, A., Murai, S., & Aoto, T. (2008). Systematic study for "Kawaii" products (the second report) -comparison of "Kawaii" colors and shapes-. In *Proceedings of SICE Annual Conference 2008* (pp. 481–484). Chofu, Japan.
6. Murai, S., Goto, S., Aoto, T., & Ohkura, M. (2008). Systematic study for "Kawaii" products (the third report) -comparison of "Kawaii" between 2D and 3D-. In *Proceedings of the 13th Annual Conference of the Virtual Reality Society of Japan* (2B5-6) (pp. 544–547). Nara, Japan. (in Japanese).
7. Ohkura, M., & Komatsu, T. (2013). Basic study on Kawaii feeling of material perception. In *Proceedings of the 15th International Conference on Human-Computer Interaction (HCII), HCII 2013* (Part I, LNCS 8004) (pp. 585–592). Las Vegas, Nevada, USA.
8. Ohkura, M., Goto, S., Higo, A., & Aoto, T. (2011). Relation between Kawaii feeling and biological signals. *Transactions of Japan Society of Kansei Engineering, 10*(2), 109–114.
9. Nittono, H. (2011). A behavioral science approach to research and development of Kawaii artifacts. *Journal of Japan Society of Kansei Engineering, 10*(2), 91–95. (in Japanese).
10. iRobot official site. Accessed August 15, 2018, https://www.irobot.com/.
11. Weber, E. H. (1978). *The sense of touch.* London, New York: Academic Press.
12. Fancher, R. E. (1996). *Pioneers of psychology.* W. W. Norton & Co.
13. Roomba 500 Series. iRobot Japan official site. Accessed August 15, 2018, https://www.irobot-jp.com/storeproduct/500series/index.html (in Japanese).
14. Sugano, S., Morita, H., & Tomiyama, K. (2012) Study on Kawaii-ness in motion -classifying Kawaii motion using roomba. In *Proceedings of AHFE 2012, The International Conference on Applied Human Factors and Ergonomics* (pp. 4498–4507). San Francisco, California, USA.
15. Sugano, S., Miyaji, Y., & Tomiyama, K. (2013). Study on Kawaii-ness in motion -physical properties of Kawaii motion of roomba. In *Proceedings of HCII 2013, The 15th International Conference on Human-Computer Interaction* S140 (Part I, LNCS 8004) (pp. 620–629). Las Vegas, Nevada, USA.

Chapter 5
Measurement and Evaluations of Kawaii Products by Biological Signals

Michiko Ohkura, Tetsuro Aoto, Kodai Ito and Ryota Horie

Abstract To evaluate the affective value of industrial products, such subjective evaluation methods as questionnaires are commonly used, even though they have some demerits such as linguistic ambiguity and interfusion of experimenter and/or participant intention to the results. We began our research to objectively evaluate interactive systems by quantifying sensations using biological signals to supplement the above questionnaire demerits. We utilize biological signals to estimate participant feelings of relaxation, comfort, and excitement, which are considered positive sensations. First, we focus on a positive and dynamic feeling called "wakuwaku." We construct various systems to evaluate affective values to derive wakuwaku feeling using biological signals and clarify the relation between the wakuwaku feeling and biological signals. Then we applied the obtained methods for wakuwaku feeling to evaluate kawaii feeling derived from various stimuli. Finally, we found that kawaii stimuli can be classified into exciting kawaii and relaxing kawaii.

Keywords Wakuwaku feeling · Kawaii · Excitement · Biological signal · Interactive system · Industrial product · Augmented reality · Virtual reality · Size · Photograph · Exciting kawaii · Relaxing kawaii

5.1 Introduction

This chapter introduces our research on kawaii feeling using biological signals mainly based on the Refs. [1–6] and their extended work.

As we have already described in Chap. 1, Kansei/affective value has become very important for industrial products. To evaluate the affective value of these products, subjective evaluation methods such as questionnaires are commonly used. In previous chapters (Chaps. 2, and 3), we introduced research on kawaii feelings of objects employing questionnaires. Questionnaires are known for the established methods

M. Ohkura (✉) · T. Aoto · K. Ito · R. Horie
Shibaura Institute of Technology, Tokyo, Japan
e-mail: ohkura@sic.shibaura-it.ac.jp

© Springer Nature Singapore Pte Ltd. 2019
M. Ohkura (ed.), *Kawaii Engineering*, Springer Series on Cultural Computing, https://doi.org/10.1007/978-981-13-7964-2_5

for subjective evaluation and have various merits. However, at the same time, they suffer from the following demerits:

- Linguistic ambiguity.
- Interfusion of experimenter and/or participant intention to the results.
- Interruption of the system's stream of information input/output.

Solving these problems is crucial to evaluate the degree of interest and/or excitement of an industrial product or an interactive system, such as whether the system is really interesting, and to identify the moment of excitement. Evaluating the affective value of an industrial product or an interactive system only by such subjective evaluation methods as questionnaires is almost impossible.

We began our research [1] to objectively evaluate interactive systems by quantifying sensations using biological signals that offer the following merits and can supplement the above questionnaire demerits:

- Can be measured by physical quantities.
- Avoids influence from the intentions of experimenter and participants.
- Can be measured continuously.

Much previous research has measured mental sensations using biological signals. Ohsuga et al. used biological signals to measure mental stress or simulator sickness [7, 8], which are considered negative sensations. On the other hand, Omori et al. measured ECG and EEG to evaluate autonomous and central nerve activities evoked by color stimuli, where they treated relaxation or comfort [9]. We previously utilized the alpha waves of EEG to estimate participant feelings of relaxation [10]. Compared with negative sensations, relaxation and comfort are considered nonnegative sensations.

In this chapter, we first focus on a feeling called "wakuwaku," which is a Japanese word for a positive sensation derived when someone feels something exciting or captivating. The word means thrilling or exhilarating in English. A wakuwaku feeling is also considered a non-negative sensation, as are relaxation and comfort. However, a big difference exists between those sensations: a wakuwaku feeling is considered dynamic, especially compared to the static sensations of relaxation and comfort. Now in 2018, much research exists on such positive and dynamic sensations as the wakuwaku feeling. However, little previous research existed in 2007 when we performed this research. In Russell's circumplex model of emotion [11] (Fig. 5.1), such negative dynamic emotions as stress is in the second quadrant, and such positive static emotions as relaxing and comfort are in the fourth quadrant. "Wakuwaku" feeling is in the first quadrant (Fig. 5.2).

The purposes of the next section include to clarify the relation between such dynamic, positive sensation as wakuwaku feeling and biological signals. Then, we introduce various kinds of research applying the obtained methods to measure kawaii feelings using biological signals, especially ECG. Finally, we found that kawaii feelings are classified into excited kawaii and relaxed kawaii. In other word, kawaii stimuli are classified into exciting kawaii and relaxing kawaii.

Fig. 5.1 Russell's
circumplex model and
locations of feelings

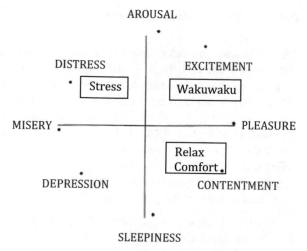

Fig. 5.2 Russell's
circumplex model and
locations of two types of
kawaii

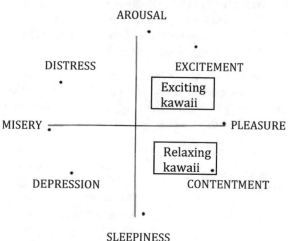

5.2 Measurement of Wakuwaku Feeling

5.2.1 Background

In this section, we focused on a feeling called "wakuwaku," a positive and dynamic sensation. The purposes of this section include to clarify the relation between wakuwaku feeling and biological signals [1].

5.2.2   *Construction of Systems*

We constructed various systems based on a treasure chest game to evaluate the
degrees of wakuwaku feeling. We employed virtual treasure boxes by computer
graphics instead of real treasure boxes because creating various real treasure boxes
mentioned below would be costly and time-consuming. These constructed systems
have various complicated components to promote wakuwaku feeling, such as the
appearances of figures, their combination, and the actions of the combined figures.
The parameters of these systems were the design of the boxes and the sound, the
BGM and the effect as shown in Table 5.1. The three box designs are shown in
Fig. 5.3. The constructed systems are shown in Table 5.2.

The procedures of the game were as follows:

1. Confirm the figures in the boxes (Fig. 5.4a).
2. Choose one of the boxes (Fig. 5.4b).
3. Watch the figure in the chosen box (Fig. 5.4c).
4. Repeat the above procedures (Fig. 5.4d).
5. Watch the combinations of the two figures (Fig. 5.4e).
6. Watch the combined figure (Fig. 5.4f).

These procedures were designed to promote wakuwaku feeling when expecting a
figure's appearance from the chosen box and combining two figures. Questionnaires
and biological signals were employed to evaluate the degree of wakuwaku feeling
of each system.

Table 5.1 System parameters

Parameters	Factor 1	Factor 2
Box design	Decorated[a]	White
Sound (BGM and effects)	With sound	Without sound

[a]Shown in Fig. 5.1

Fig. 5.3 Three types of boxes

Table 5.2 Constructed systems

System	Box design	Sound
1	Decorated	With sound
2	Decorated	Without sound
3	White	With sound
4	White	Without sound

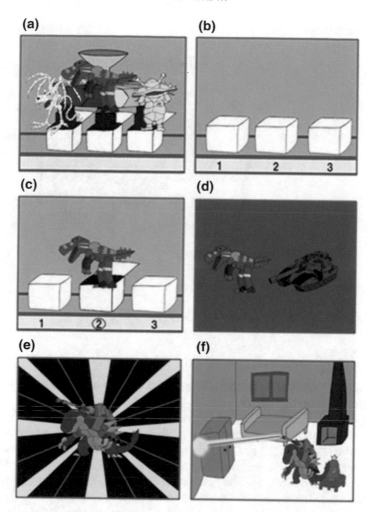

Fig. 5.4   System flow

Figure 5.5 shows the system setup. The input device was a keypad, and the output devices were a 17-inch LCD display and a pair of speakers. The biological signals were measured by sensors and BIOPAC measurement equipment (Biopac Systems Inc.). Two PCs were employed for system display and to measure the biological signals.

Fig. 5.5 System setup

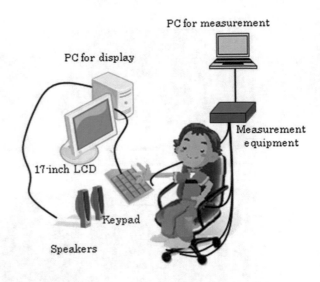

5.2.3   Experiments to Evaluate the Systems

5.2.3.1   Method

The participants randomly played the four games shown in Table 5.2 and answered questionnaires for each system.

The questionnaire about the wakuwaku feeling consisted of 23 items of paired 7-point evaluations such as "fun boring," and five items of unpaired 5-scale evaluation such as "pounding." For the paired 7-point evaluation items, four indicates neutral, seven indicates the best, and one indicates the worst. For the unpaired 5-point evaluation items, five indicates the best and one the worst. In addition, participants were asked some free description questions after playing four games.

The following biological signals were measured constantly during the experiments to detect the degree of wakuwaku feeling: Galvanic Skin Reflex (GSR), Electrocardiogram (ECG), and breathing rate. GSR, which is affected by states of emotion, was used as a physiological index to detect such emotions as anxiety and mental stress. ECG is changed not only by physical exercise but also by mental factors such as anxiety and stress. In addition, breathing rates and patterns are also indexes of stress and anxiety.

5.2.3.2   Results

Experiments were performed with 12 male students in their 20s who served as volunteers.

From the results of the analysis of variance for each questionnaire item with parameters in Table 5.1, the main effect of sound was significant for almost all questionnaire items including exciting and enjoyable. On the other hand, the main effect of the box design was not significant for almost all items. Table 5.3 shows the result of the analysis of variance for enjoyable. In the free description answers, some participants pointed out that the BGM and the sound effects were good points of the system, suggesting that sound is effective for wakuwaku feelings.

As for biological signals, we selected various physiological indexes as shown in Table 5.4. The RR interval, defined as the time interval between the two R waves of the ECG, is the inverse of the heart rate, the number of heart beats per minute. All indexes were normalized by the values in the quiet state for each participant. Since we designed the game flow with various events to promote wakuwaku feelings, we chose the following three moments for analysis:

- Moment I: When the first box opened.
- Moment II: When the second box opened.
- Moment III: Just after combining the two figures.

The first and second moments were in the first half of the game, while the third moment was in the second part. By a paired difference test, the heart rate at each moment of the first and the second choices of boxes was significantly different

Table 5.3 Analysis of variance (enjoyable)

Factor	Sum of squared deviation	DOF	Mean square	F-value	P-value
Box design	20.02	1	20.02	14.47	0.00**
Sound	0.19	1	0.19	0.14	0.71
Error	62.27	45	1.38		
Total	82.48	47			

**: Significant at the 1% level

Table 5.4 Physiological indexes

Biological signal	Physiological indexes
Galvanic Skin Reflex (GSR)	Average GSR
Electrocardiogram (ECG)	Average heart rate (the average number of heart beats per minute)
	Variance of heart rate
	Average RR interval (inverse of heart rate)
	Variance of RR interval
Breathing rate	Number of breaths (breath per minute)
	Variance of number of breaths
	Amplitude of breathing

Table 5.5 Results of difference tests

Physiological index	Parameter	Moment I	Moment II	Moment III
Average GSR	Box design	–	–	–
Average GSR	Sound	–	–	*
Average heart rate	Box design	**	**	–
Average heart rate	Sound	–	–	–
Average RR interval	Box design	**	**	–
Average RR interval	Sound	–	–	–

– : Not significant, *: Significant at the 5% level, **: Significant at the 1% level

between the systems with different box designs. On the other hand, the heart rate at Moments I or II was not significantly different between the systems with sound and without sound. However, the averages of GSR at Moment III were significantly different only between the systems with sound and without sound. Table 5.5 summarizes the results of all tests.

5.2.4 Discussion

The above experimental results suggest that heart rate and GSR averages may show the wakuwaku feeling of the users of interactive systems. Moreover, the heart rate results are related to the system's former part, and the results of GSR averages are related to its latter part. Since the questionnaire results agreed with the results of the GSR averages and disagreed with the heart rate results, they might reflect the wakuwaku feeling of the latter part of the systems. The questionnaire may reflect the wakuwaku feeling of the system's last part because the participants can only remember it.

5.2.5 Summary

To evaluate the affective value of industrial products, we constructed various systems based on a treasure chest game to evaluate their affective values especially generated wakuwaku feeling using biological signals.

We performed experiments to measure the degree of wakuwaku feeling by using the constructed systems. From analysis of the experimental results, we obtained the following useful knowledge:

- The degree of wakuwaku feeling may vary depending on such parameters as object design and sound effects.
- The degree of wakuwaku feeling may be measured by such biological signals as GSR and ECG.

This work is the first step to measure wakuwaku feeling that occurred by interactive systems. We performed many researches on wakuwaku feeling after that such as [12–16].

5.3 Measurement of Kawaii Feeling and Biological Signals—Kawaii Color and Size

5.3.1 Background

From 2007, we focused on the kawaii attributes of industrial products, because we considered kawaii one of the most important affective values. Our aim is to clarify a method for constructing kawaii products from the research results. We previously performed experiments and obtained valuable knowledge on kawaii attributes (Chap. 2). For example, curved shapes such as a torus and a sphere are generally evaluated as more kawaii than straight-lined shapes. Brightness and saturation are effective for kawaii colors.

Meanwhile, only questionnaires were employed to evaluate kawaii shapes and colors in all the above experiments. However, questionnaires suffer from some demerits as described in Sect. 5.1. Thus, to compensate for questionnaire demerits, we examined the possibility of using biological signals.

This section describes our trials to clarify the relation between kawaii colors and biological signals and the relation between kawaii sizes and biological signals [2].

5.3.2 Kawaii Colors and Biological Signals

5.3.2.1 Experimental Method

The first experiment addressed kawaii colors. We performed a preliminary experiment to select color candidates from 381 colors in the color table [17] employing four male and two female students in their 20s as participants. As the experimental results, pink and colors basically identical to pink were selected as kawaii colors, while dark brown and dark green were selected as non-kawaii colors. Thus, we selected pink (5R 7/10 from the Munsell Color System [18], blue (10B 7/6), brown (5R 4/6), and green (between 2G 4/4 and 3G 4/4) for the following evaluation experiment, where green represented a middle color between kawaii and non-kawaii colors.

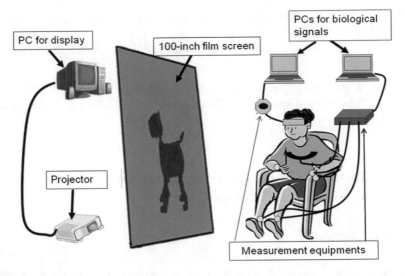

Fig. 5.6 Experimental setup

Figure 5.6 shows the experimental setup. Participants watched a large, 100-inch screen whose surface was covered by one of the four above colors and evaluated its kawaii degree on a 7-scale evaluation.

The experimental procedures were as follows:

1. Participants sat on chairs.
2. The experimenter explained the experiment.
3. Participants wore electrodes.
4. Participants remained quiet for 30 s.
5. Participants watched the screen displayed by a color for 30 s.
6. Participants answered the questionnaire.

Steps 5 and 6 were repeated four times for the four colors, which were displayed in random order. Such biological signals as heart rate, Galvanic Skin Reflex (GSR), breathing rate, and Electroencephalogram (EEG) were measured both before and while watching. The biological signals were measured by BIOPAC Student Lab (BIOPAC Systems, Inc.), except for EEG which was measured by the Brain Builder Unit (Brain Function Research Center).

5.3.2.2 Experimental Results

Experiments were performed with eight female and eight male student volunteers in their 20s. Based on our previous experiments [19–22], we selected the following physiological indexes for analysis:

- Average heart rate,
- Variance of heart rate,
- Average RR interval,
- Variance of RR interval,
- Average GSR,
- Variance of GSR,
- Number of breaths,
- Variance of number of breaths,
- Average breath magnitude,
- Variance of breath magnitude,
- Ratio of power spectrum of Theta, Slow alpha, Mid alpha, Fast alpha, and Beta waves,
- Ratio of dominant duration of Theta, Slow alpha, Mid alpha, Fast alpha, and Beta waves.

All indexes described above were normalized by the values in the quiet state for each participant.

Figure 5.7 shows the questionnaire results, where the horizontal axis shows the participants (male: a–h, female: i–p) and the vertical axis shows the kawaii degree of each color.

From the results of a two-factor analysis of variance, the main effect of color is significant at the 1% level, the main effect of gender is significant at the 5% level, and a significant cross-effect exists between color and gender at the 5% level. Thus,

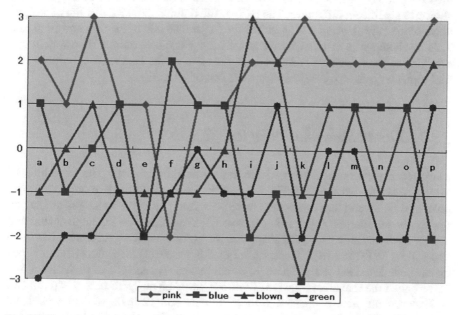

Fig. 5.7 Questionnaire results of the first experiment

Table 5.6 Unpaired t-test of difference of mean value of heart rate

Factor	Kawaii	Non-kawaii	Difference	Equality of variance	T-Test
Number	32	26		F:1.74	T:1.75
Average	3.14	0.88	2.26	DOF1:25	DOF:56
Unbiased variance	17.92	31.11		DOF2:31	0.08
SD	4.23	5.58	1.34	P:0.15	0.04*

*: Significant at the 5% level

we successfully analyzed the biological signals by dividing the participants into two groups by kawaii scores.

The data of the biological indexes were divided into the following two groups: the kawaii group where scores were above 0, and the non-kawaii group where the kawaii scores were below 0. The data with 0 score were omitted from analysis. From the unpaired t-test results of the difference of the mean value of the two groups, the heart rates, the numbers of heart beats per minute, showed a significant difference (Table 5.6). The heart rate of the kawaii group was significantly faster than that of the non-kawaii group. This result suggests that watching a kawaii color is more exciting than watching a non-kawaii color, because a decrease in the heart rate is considered an index of relaxation.

Moreover, from the unpaired t-test results of the difference of the mean value of the two groups by gender, the ratio of the dominant duration of the mid-alpha wave showed a significant difference. The ratio of the dominant duration of the mid-alpha wave of the kawaii group was significantly larger than that of the non-kawaii group. This result suggests that watching a kawaii color is more exciting than watching a non-kawaii color, because a decrease of the ratio of the dominant duration of the mid-alpha wave is also considered as an index of relaxation.

5.3.3 Kawaii Sizes and Biological Signals

The second experiment addressed the kawaii sizes of objects. The experimental setup resembled that shown in Fig. 5.4, except the display used two projectors with polarized filters and the participants wore polarized glasses to watch the objects on the screen stereoscopically. Participants watched an object on the large screen and evaluated its kawaii degree. The shape and the color of each object were set as torus and yellow, 5Y8/14 in MCS, as shown in Fig. 5.8, based on the results of our previous experiment described above. The size, which means visual angle, of each object was set as one of the four sizes shown in Table 5.7 based on our preliminary experiment.

Experiments were conducted with 12 female and 12 male student volunteers in their 20s. Figure 5.9 shows the questionnaire results, where the horizontal axis shows the participants (male: m1–m12, female: f1–f12) and the vertical axis shows

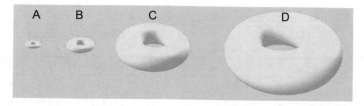

Fig. 5.8 Objects employed for second experiment

Table 5.7 Analysis of variance

Object	A	B	C	D
SIZE	1	2	6	10
Vertical degree	3.7	7.5	22.0	35.4
Horizontal degree	5.0	10.2	29.5	47.3

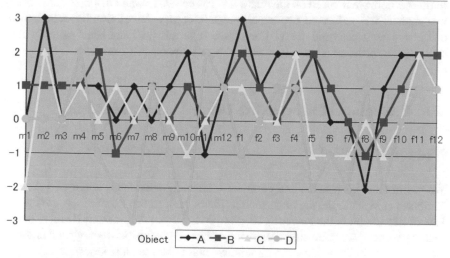

Fig. 5.9 Questionnaire results of second experiment

the kawaii degree for each object size. The results of a two-factor analysis of variance show that the main effect of size is significant at the 1% level and the main effect of gender is not significant. The biological signal data were divided into two groups similar to the first experiment. From the unpaired t-test results of the difference of the mean value of the two groups, the heart rates showed a significant difference (Table 5.8). Since a higher heart rate shows the unrelaxed state, the mental state when feeling kawaii is considered more exciting than not feeling kawaii. In addition, the results of the similar difference test of the heart rate for object C showed a significant difference at the 5% level, and the results of the similar difference test of the heart rate for object D showed a significant difference at the 1% level. These results show that the mental state when feeling kawaii is probably more exciting than not feeling kawaii even if the size of the object being watched is the same.

Table 5.8 Unpaired t-test of difference of mean value of heart rate

Factor	Kawaii	Non-kawaii	Difference	Equality of variance	T-test
Number	50	20		F:2.99	T:2.555
Average	3.45	0.60	2.85	DOF1:49	DOF:68
Unbiased variance	18.98	14.61		DOF2:19	0.013
SD	4.36	3.82	0.534	P:0.542	0.006**

**: Significant at the 1% level

5.3.4 Summary

In this section, we focused on the relation between two kawaii attributes and biological signals. We performed two experiments to clarify the relation between kawaii colors and heart rates and the relation between kawaii sizes and heart rates. In both experiments, heart rates increased when the participants felt kawaii.

5.4 Evaluation of Kawaii Size Using Augmented Reality

5.4.1 Background

We performed experiments and obtained interesting tendencies about such kawaii attributes as sizes in the previous section (Sect. 5.3). In this section, we refocus on kawaii size to obtain the tendency of kawaii sizes using Augmented Reality. In our first experiment, we compared the results of virtual objects between virtual and real environments. In the second experiment, we classified the most kawaii size.

5.4.2 The First Experiment

5.4.2.1 Experimental Method

In our experimental setup, we employed a 42-inch LCD 3D monitor (Hyundai Corp.) and polarized glasses to stereoscopically show virtual objects in a virtual environment (VR) (Fig. 5.10). The distance between the monitor and each participant was 1.0 m. In a real environment, we employed 3D see-through glasses (Wrap920AR, Vizix Corporation) to stereoscopically show the virtual objects (AR) (Fig. 5.11). The marker for the AR display was set in the left hand of each participant. The distance between the participants and their hands was 0.5 m.

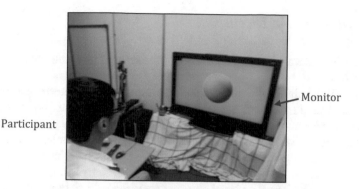

Participant

Monitor

Fig. 5.10 VR setup

Participant

Marker

Fig. 5.11 AR setup

Table 5.9 Visual angles of virtual objects both in VR and AR

Number	1	2	3	4
Visual angle (°)	2.3	4.6	9.1	18.2

Participants watched an object in VR or AR and evaluated its kawaii degree. Each object's shape was set as a cube and its color was set as red-yellow (orange) on the basis of the results of our previous experiments (Chap. 2). Each object's size, which implies a visual angle, was set to one of four (Table 5.9) both in VR and AR based on our previous experiment (Sect. 5.3). We randomly showed these four objects to the participants, who evaluated the kawaii degree for each of them on a 7-point Likert scale:-3: extremely not kawaii, 0: neutral, and +3: extremely kawaii. We also measured the ECGs of the participants during the experiment.

5.4.3 Results and Discussion

Our experiments were performed with eight female and eight male student volunteers in their 20s.

Figure 5.12 exhibits the questionnaire results, where the horizontal axis shows the ratios of the kawaii scores and the vertical axis shows each object size in VR. The averaged scores for the kawaii degrees show that smaller objects tend to be more kawaii in VR, which is the same result as in a previous experiment (Sect. 5.3). The results of a two-factor analysis of variance for VR show that the main effect of size is significant at the 1% level, but the main effect of gender is not.

Figure 5.13 displays the virtual objects in AR, and Fig. 5.14 shows the questionnaire results for AR. The averaged scores for the kawaii degrees in AR show the same results in VR. The results of a two-factor analysis of variance for AR show that the main effect of size is significant at the 1% level, but the main effect of gender is not.

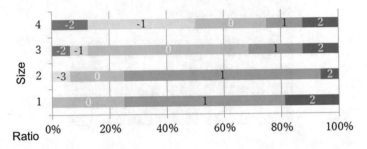

Fig. 5.12 Questionnaire results for VR

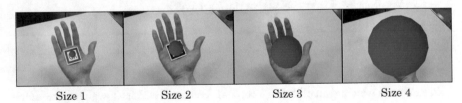

Size 1   Size 2   Size 3   Size 4

Fig. 5.13 Displays of virtual objects in AR

Fig. 5.14 Questionnaire results for AR

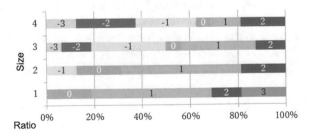

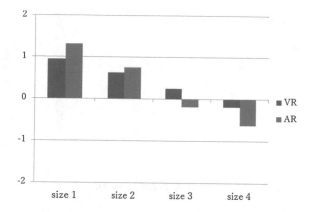

Fig. 5.15 Comparison of averaged kawaii degrees between VR and AR

Figure 5.15 compares the averaged kawaii degrees for each size between VR and AR, where the vertical axis shows the averaged kawaii degree. The correlation coefficients between the sizes of the objects and the kawaii degrees are higher in AR than VR, implying that the AR result is stronger than that of VR.

We calculated heart rates from ECG signals. Heart rates were normalized by the values in the quiet state for each participant. The data of the biological indexes were divided into the following two groups: the kawaii group where scores were above 0 and the non-kawaii group where the kawaii scores were below 0. The data with 0 score were omitted from analysis. From the unpaired t-test results of the difference of the mean value of the two groups, the heart rates, the numbers of heart beats per minute, showed a significant difference only in case of AR. The heart rate of the kawaii group was significantly faster than that of the non-kawaii group in case of AR. This result reconfirmed that kawaii feeling increases the heart rate.

Table 5.10 compared the averaged heart rates between feeling kawaii and non-kawaii in VR and AR. Figure 5.16 showed the averaged heart rates for each size in VR and AR. This figure shows the strong correlation between size and heart rate in AR.

Table 5.10 Comparison of the averaged heart rate between feeling kawaii and non-kawaii in VR and AR

		VR	AR	Average
Heart rate	Kawaii	2.36	3.23	2.795
	Non-kawaii	−0.91	0.90	−0.01

Fig. 5.16 Averaged heart
rate for each size in VR and
AR

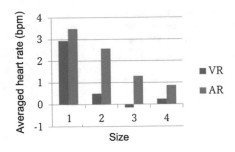

5.4.4 The Second Experiment

5.4.4.1 Experimental Method

The results of the first experiment showed that participants tended to judge smaller objects more kawaii. However, since the most kawaii size was not clarified, we did the second experiment.

Both for AR and VR, we employed the same setup as in the first experiment. The initial size of the virtual object was set as the smallest, as in Table 5.1, and participants could increase or decrease its size using the arrow key of the keyboard until they felt the object to be the most kawaii. The pitch of the change of the object's size was 0.23°, which was 10% of the initial size.

5.4.4.2 Results and Discussion

The second experiment was performed with a female and four male student volunteers in their 20s.

Figure 5.17 shows the results, where the initial size was set to 1 on the vertical axis. Participant A should be considered as an outlier, because the most kawaii size for the other participants almost equaled double the object's initial size. This result shows that the most kawaii size has a lower limit.

The results of our first experiment suggested that smaller objects tend to be more kawaii. However, our second experiment results showed that the tendency of size of

Fig. 5.17 The most kawaii
size

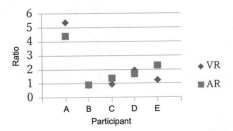

the first experiment was not always, and that an optimum size for kawaii objects was determined.

5.4.5   Summary

We performed two experiments to determine the most kawaii size using both VR and AR. In the first experiment, our results suggest that smaller objects are more kawaii in both environments. In addition, the AR result is stronger than that of VR from both questionnaires and heart rates. In the second experiment, we determined for kawaii objects an optimum size, implying a limit on the tendency obtained in our first experiment.

5.5   Evaluation of Kawaii Objects in Augmented Reality by ECG

5.5.1   Background

In this section, we employed ECGs in our evaluation to clarify the differences of kawaii feelings between different objects using augmented reality (AR) based on the previous section (Sect. 5.4).

5.5.2   Experimental Method

In our experimental setup, we employed 3D see-through glasses to stereoscopically show virtual objects. We set markers for the AR display in the left hands of the participants, who observed three objects in AR. One was an orange ball employed in our previous research (Fig. 5.18a). Another was chosen by each participant as the most ordinary object from the five candidates shown in Fig. 5.18b. The last object was also chosen by each participant as the most kawaii object from the five candidates in Fig. 5.18c.

We measured the ECGs before and while observing the objects. Because of the individual differences of heart rates (HRs), we used the difference between the HR while and before observing the objects as normalized HRs.

Fig. 5.18 Objects observed
by participants

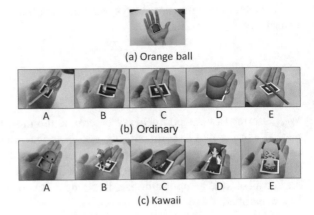

(a) Orange ball

A   B   C   D   E

(b) Ordinary

A   B   C   D   E

(c) Kawaii

Fig. 5.19 Normalized HRs
for all participants

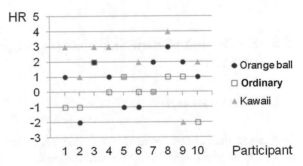

5.5.3 Experimental Results and Discussion

We performed our experiments with 5 female (1–5) and 5 male (6–10) student volunteers in their 20s. Figure 5.19 compares their averaged normalized HRs.

When they observed the kawaii objects, the HRs were high for all of the females
and most of the males. When observing the common objects, the HRs resembled the
HRs before observing the objects for all the females and most of the males.

5.5.4 Summary

We experimentally clarified the relation between objects and ECGs. We observed
kawaii objects in AR related to high heart rates and got the same results as from the
elderly who observed kawaii spoons (Chap. 8).

5.6 Differences in Heartbeat Modulation Between Excited and Relaxed Kawaii Feelings During Photograph Observation

5.6.1 Background

A person experiences kawaii feelings when observing kawaii objects or attributes in everyday sceneries. Specifically, watching photographs elicits subjective, intuitive, and strong preferences depending on personal emotional feelings, suggesting that investigation of kawaii feeling that occurs while watching photographs provides valuable insights. Therefore, kawaii-related physiological changes while performing this task were measured. In a previous study, we detected heartbeat modulation when participants experienced kawaii feeling watching photographs [23, 24]. A pronounced modulation was observed when the subjects themselves selected kawaii photos that act as stimuli. Moreover, physiological brain responses were measured in event-related potentials when participants experienced kawaii feeling when watching photographs [25, 26].

We further hypothesized that the kawaii feeling evoked while watching photographs could be classified into two types [27]. The first type, called "excited kawaii" in this section, corresponds to a feeling of cuteness accompanied by excitation. The second type, or "relaxed kawaii," is a feeling of cuteness accompanied by relaxation. To test this hypothesis, in this study, heartbeat modulations were measured when participants experienced both types of kawaii feeling while watching photographs and potential differences in these heartbeat modulations were examined with respect to both types.

This section is described based on Ref. [4].

5.6.2 Material and Methods

5.6.2.1 Participants

Seven healthy students in their 20s participated in the experiment. All participants gave informed consent prior to participation. The experimental procedure was approved by the ethics committee of Shibaura Institute of Technology.

5.6.2.2 Stimuli

Each participant subjectively selected (1) an exciting kawaii photograph, which evoked excited kawaii feeling, (2) a relaxing kawaii photograph, which evoked relaxed kawaii feeling, and (3) an uninteresting photograph, which evoked no specific feeling. Exciting kawaii photographs were expected to make the participants' hearts palpitate and in turn enhance their heart rates. Relaxing kawaii photographs were assumed to make the participants unwind, resulting in lower heart rate increase compared with their exciting kawaii counterparts. Uninteresting photographs were not expected to cause any heartbeat modulation and acted as control stimuli. Heartbeat modulation differences according to the types of photographs would physiologically verify the hypothesis that the kawaii feeling resulting from watching photographs can be categorized as excited and relaxed.

Participants were instructed to select photographs that induced the desired feelings in a stable and permanent manner from the Internet. Typical exciting kawaii photographs showed animals and portraits across the subjects while their relaxing kawaii counterparts usually displayed animals. Uninteresting photographs generally presented tools, furniture, buildings, or sceneries. Interestingly, one participant selected an animal picture as an exciting kawaii photograph that was chosen as a relaxing kawaii one by another participant. This suggested that the kawaii classification strongly depended on subjective feelings. Selected photographs were converted to monochrome images and edited to fit a 192 pixel-high by 256 pixel-wide frame.

5.6.2.3 Heart Rate Measurement

Visual stimuli were displayed on a laptop computer screen placed in front of the participants (Fig. 5.20). During each trial, participants quietly faced a black screen for 30 s (rest period) before watching a picture for 30 s (task period). This trial was repeated five times for a total duration of 300 s while an electrocardiogram (ECG) was measured by radiofrequency ECG (Wireless Vital Sign Sensor, Micro Medical Device, Inc.). Heart rates were calculated from RR intervals in individual ECGs.

The period taken to present each picture was longer that than in previous experiments [23] in order to provide sufficient time to significantly change heart rates.

Fig. 5.20 A snapshot of the experiment being conducted

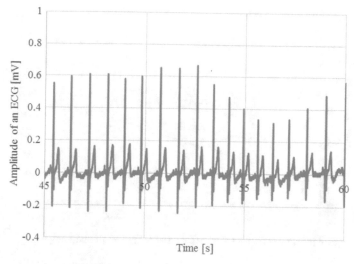

Fig. 5.21 A typical waveform of the heartbeat

The time interval between photograph selection and ECG measurement was limited to 10–15 min to ensure that the participant's impressions of the photos remained unchanged.

5.6.3   Results

5.6.3.1   ECG Waveforms

A typical ECG waveform (Fig. 5.21) shows the amplitude [mV] as a function of time [s] within a trial. This waveform was extracted from a trial using an exciting kawaii photograph. Figures 5.22, 5.23, and 5.24 show typical heart rates obtained during each trial for a participant using exciting kawaii, relaxing kawaii, and uninteresting photographs, respectively. In Figs. 5.22, 5.23, and 5.24, numbers in legends are corresponding to the trials. Horizontal axes indicate the rest and task periods while vertical axes represent heart rates in a trial [bpm]. Heart rate distributions are shown for each trial.

5.6.3.2   Average Heart Rate Difference

For each trial, a heartbeat modulation evoked by a photograph was defined as a change in heart rate average during the second half of the task period (15 s), which was assumed as a baseline, relative to the heart rate average during the first half of the task period (15 s). This change was called heart rate difference. For each

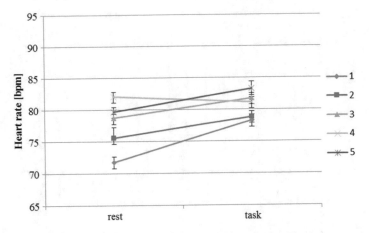

Fig. 5.22 Typical heart rates for each trial using an exciting kawaii photograph

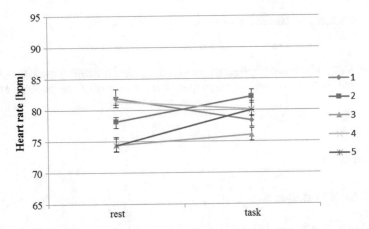

Fig. 5.23 Typical heart rates for each trial using a relaxing kawaii photograph

participant, five heart rate differences were obtained and averaged for the five trials conducted using individual pictures. These averages were defined as average heart rate differences (Table 5.11).

5.6.3.3 Statistical Tests

Statistical tests were conducted using average heart rate differences (Table 5.11). To determine whether a type of photograph induced a heartbeat modulation, a t-test was performed for these average differences according to the stimulus. Exciting kawaii photographs produced a significant difference ($p < 0.05$), whereas relaxing kawaii and uninteresting photographs induced little changes.

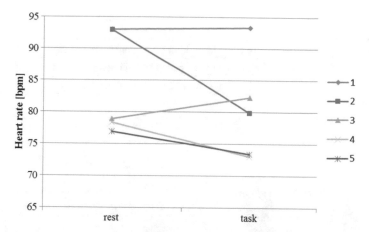

Fig. 5.24 Typical heart rates for each trial using an uninteresting photograph

Table 5.11 Average heart rate differences for each type of photograph and individual subject	Participants	Exciting kawaii photo	Relaxing kawaii photo	Uninteresting photo
	A	3.063	1.266	−3.623
	B	2.164	3.620	−0.864
	C	−0.960	−1.919	−2.700
	D	0.713	0.339	1.619
	E	1.695	1.612	0.555
	F	4.003	−0.711	−0.764
	G	2.155	1.261	−1.580
	Average	1.833	0.781	−1.051
	p-value	0.024	0.289	0.173

ANOVA was conducted to evaluate whether heartbeat modulation depended on the type of photograph presented. A significant difference was observed between the three stimuli ($p = 0.044$). Table 5.12 shows post hoc multiple comparison results of the one-way ANOVA using the Bonferroni correction. A significant difference was obtained between exciting kawaii and uninteresting photographs ($p < 0.05$). No significant differences were detected between exciting and relaxing kawaii photographs or between relaxing kawaii and uninteresting photographs. Figure 5.25 shows one-way ANOVA and post hoc multiple comparison results. Average heart rate differences [bpm] are displayed for individual exciting kawaii, relaxing kawaii, and uninteresting photographs. Average heart rate difference distributions are also presented for the seven participants.

Table 5.12 Post hoc multiple comparison results of the one-way ANOVA

	p-value
Exciting kawaii photo relaxing kawaii photo	0.811
Exciting kawaii photo and uninteresting photo	0.018
Relaxing kawaii photo and uninteresting photo	0.189

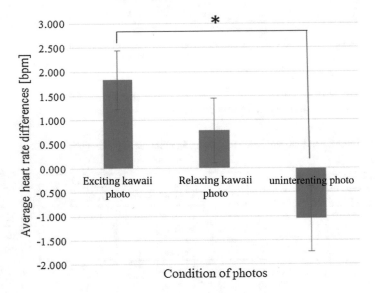

Fig. 5.25 A typical waveform of the heartbeat

5.6.4 Discussion

A significant t-test result was observed only for exciting kawaii photographs, indicating that these photographs increased heart rates. Moreover, significant differences were detected between exciting kawaii and uninteresting photographs, whereas relaxing kawaii and uninteresting photographs produced little changes. These results validate the hypothesis that the kawaii feeling evoked while watching photographs can be classified as excited and relaxed kawaii feelings. Exciting kawaii photographs produced a heartbeat modulation, unlike their relaxing kawaii counterparts.

The heart rate increase induced by the exciting kawaii photographs was consistent with previous observations for participants watching kawaii pictures [23]. Here, relaxing kawaii photographs did not boost heart rates to the same extent as their exciting kawaii counterparts, showing that the relaxing effects of healing photographs may suppress the increase in heart rate [28]. Furthermore, excited and relaxed kawaii feelings may correspond to two opposite directions on the arousal axis of Russell's circumplex model of emotion.

5.6.5 Summary

This section verified the hypothesis that the kawaii feelings evoked while watching photographs can be classified as excited and relaxed kawaii feelings. Heartbeat modulations occurring in participants experiencing these feelings while watching photographs were measured. Exciting kawaii photographs significantly enhanced heart rates, whereas their relaxing kawaii counterparts induced little heart rate increases, validating our hypothesis. Overall, the heartbeat responses resulting from watching kawaii photographs exhibit different modulations depending on the two types of kawaii feeling, although only one word is used in our daily lives.

5.7 Detection of Relaxing Kawaii Using Stuffed Animals and ECG

5.7.1 Background

Our previous studies revealed that the physiological responses evoked by kawaii feelings can be measured by ECG (Sects. 5.3–5.5). We also verified that exciting kawaii photographs significantly enhanced heart rates, while their relaxing counterparts did not have this effect (Sect. 5.6). Exciting kawaii response is located in the first quadrant of Russell's circumplex model, and that of relaxing kawaii is in its fourth quadrant. In this study, we employed stuffed animals as stimuli for relaxing kawaii and clarified the resulting effects using ECG measurements [5].

5.7.2 Experimental Method

The task was to look at six stuffed animals one by one. Pulse waves were measured as participants did this looking task. We used a pulse wave sensor that can measure heartbeat by simply attaching it to the left index finger, which lessens the psychological burden on participants. The sensor was connected to an amplifier (NEXUS-10 Mark II) to send the output to a PC. Consequently, heartbeat could be monitored in real time. Figure 5.26 shows our experimental system.

The participant was a female in her forties. A two-stage procedure of a 30-s rest period, during which the participant does not look at anything, and a 30-s task period, during which she looks at a single stuffed animal, was repeated six times. Three out of the six stuffed animals were owned by the participant, but she had never seen the other three before. Figure 5.27 shows the stuffed animals used in the task and their presentation order.

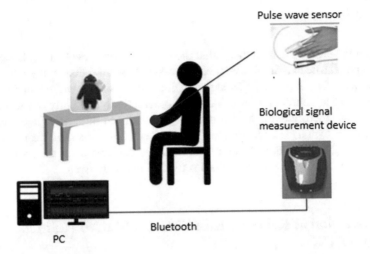

Pulse wave sensor

Biological signal
measurement device

Bluetooth

PC

Fig. 5.26 Experimental system

Fig. 5.27 Stuffed animals in the task

5.7.3 Experimental Results

We calculated the participant's heart rate at rest for 30s and during the 30s tasks based on the RR intervals of the pulse wave. Figure 5.28 shows an experimental scene. Figure 5.29 shows the heart rate calculated for each trial, where the vertical axis shows the heart rate and the horizontal axis shows the rest and task periods in each trial. As shown in the figure, the heart rate was lower in most trials while the participant looked at the stuffed animals than in the resting state. This is an important finding because the heart rate showed that the participant felt super-relaxing kawaii while looking at stuffed animals.

5.7.4 Summary

We performed an experiment using stuffed animals and ECG measurement. Calculated heart rates showed that the participant felt relaxing kawaii while looking at stuffed animals.

Fig. 5.28 Experiment scene

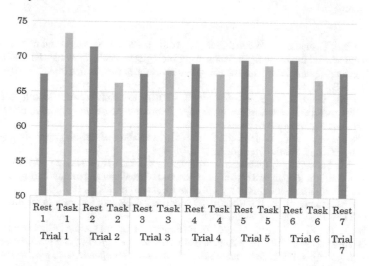

Fig. 5.29 Calculated heart rate for each trial

In the previous section (Sect. 5.6), the relaxing kawaii photograph gave no significant change from resting state. Therefore, the success of the detection of relaxing kawaii by heart rates decreases in this section is an important new finding.

5.8   Conclusions

This chapter described various researches on measurement of kawaii feeling using biological signals. In Sect. 5.1, we introduced the merits of using biological signals to measure positive dynamic emotions. In Sect. 5.2, we introduced the measurement of wakuwaku feeling. In Sects. 5.3–5.5, we introduced the measurement of exciting kawaii feelings by color, size, and visual stimulus using ECG. In Sects. 5.6 and 5.7, we introduced that kawaii feeling can be classified into excited kawaii and relaxed kawaii.

Acknowledgements This research was partly supported by the Grant-in-Aid for Scientific Research (C) (No.17500150), (C) (No. 21500204), (B) (No. 2628104), and (C) (No. 17K00388), Japan Society for the Promotion of Science, and by the SIT Research Promotion Funds. We thank the students of Shibaura Institute of Technology who contributed to this research and served as volunteers.
In addition, we thank the staff of Tokyo Broadcasting System Television, Inc. (TBS) for their help in the experiment described in Sect. 5.7.

References

1. Ohkura, M., Hamano, M., Watanabe, H., & Aoto, T. (2010). Measurement of "wakuwaku" feeling generated by interactive systems using biological signals. In *Proceedings of the Kansei Engineering and Emotion Research International Conference 2010 (KEER 2010), Paris* (pp. 2293–2301).
2. Ohkura, M., Goto, S., Higo, A., & Aoto, T. (2010). Relation between kawaii feeling and biological signals. In *Proceedings of the 3rd International Conference of Applied Human Factors and Ergonomics 2010, Miami*.
3. Ohkura, M., Yamasaki, Y., & Horie, R. (2014). Evaluation of kawaii objects in augmented reality by ECG. In *Proceedings of EMBC 2014*.
4. Yanagi, M., Yamariku, Y., Takashina, T., Hirayama, Y., Horie, R., & Ohkura, M. (2016). Differences in heartbeat modulation between excited and relaxed kawaii feelings during photograph observation. *International Journal of Affective Engineering, 15*(2), 189–193.
5. Ohkura, M., Tombe, T., & Ito, K. (2019). Detection of relaxing kawaii using stuffed animals and ECG. In *Proceedings of EMBC 2018*.
6. Tombe, T., Ito, K., & Ohkura, M. (2019). Evaluation of kawaii feelings caused by looking at stuffed animals using ECG. In S. Fukuda (Ed.), *Advances in affective and pleasurable design-proceedings of the AHFE 2018 international conference on affective and pleasurable design* (pp. 386–391). Cham: Springer.
7. Hirao, N., et al. (1997). Assessment of mental stress during a cooperative work using stress detection method triggered by physiological data. *ITE Technical Report, 21,* 67–72. (in Japanese).
8. Sakaguchi, T., et al. (1999). Stress detection in a virtual ensemble using autonomic indices. In *Proceedings of the 59th National Convention of IPSJ* (pp. 2-1–2-2) (in Japanese).
9. Omori, M., Hashimoto, R., & Kato, Y. (2002). Relation between psychological and physiological responses on color stimulus. *Journal of the Color Science Association of Japan, 26*(2), 50–63. (in Japanese).
10. Ohkura, M., & Oishi, M. (2006). An alpha-wave-based motion control system for a mechanical pet. *Kansei Engineering International, 6*(2), 29–34.
11. Russell, J. (1980). A circumplex model of affect. *Journal of Personality and Social Psychology, 39,* 1161–1178.
12. Harada, Y., Furuya, T., Takahashi, N., Hasegawa, K., Nakazato, T., & Ohkura, M. (2014). Content evaluation of exciting feeling by using biosignals. In *Proceedings of AHFE 2014* (pp. 6931–6936).
13. Harada, Y., Nakatsuji, H., Tate, Y., Seto, H., & Ohkura, M. (2015). Evaluation of exciting feeling in driving simulation by HRV. In *Proceedings of JSST2015*.
14. Ito, K., Harada, Y., Tani, T., Hasegawa, Y., Nakatsuji, H., & Tate, Y., et al. (2016). Evaluation of feelings of excitement caused by auditory stimulus in driving simulator using biosignals. In *Proceedings of APD 2016* (pp. 231–240).

15. Ito, K., Usuda, S., Yasunaga, K., & Ohkura, M. (2017). Evaluation of "feeling of excitement" caused by a VR interactive system with unknown experience using ECG. In *Proceedings of APD 2017* (pp. 292–304).
16. Ito, K., Miura, N., & Ohkura, M. (2019). Affective evaluation of a VR animation by physiological indexes calculated from ECGs. In *Proceedings of IEA2018, AISC 827* (pp. 329–339).
17. Azo (2010). Color guide. http://www.color-guide.com/e_index.shtml.
18. Emori, Y., et al. (1979). *Colors–their science and culture–*. Tokyo: Asakura Publishing Co., Ltd. (in Japanese).
19. Aoki, Y., et al. (2006). Assessment of relief level of living space using immersive space simulator (Report 4)–Examination of physiological indices to detect "Uneasiness". In *Proceedings of VRSJ the 10th Annual Conference, Tokyo, Japan, (CD-ROM), 2B1-4*. (in Japanese).
20. Ohkura, M., et al. (2009). Evaluation of comfortable spaces for women using virtual environment-objective evaluation by biological signals. *Kansei Engineering International*, *8*(1), 67–72.
21. Aoto, T., & Ohkura, M. (2007). Study on usage of biological signal to evaluate kansei of a system. In *Proceedings of the 9th Annual Conference of JSKE 2007, Tokyo, Japan, (CD-ROM), H-27* (in Japanese).
22. Aoto, T., & Ohkura, M. (2007). Study on usage of biological signal to evaluate kansei of a system. In *Proceedings of the 1st International Conference on Kansei Engineering and Emotion Research (KEER2007), Sapporo, Japan, (CD-ROM), L-9*.
23. Yanagi, M., Yamasaki, Y., Yamariku, Y., Takashina, T., Hirayama, Y., & Horie, R., et al. (2013). Relation between kawaii feeling and heart rates in watching photos. In *Proceedings of the 15th Japan Society of Kansei Engineering Conference and General Meeting B* (Vol. 12). Tokyo (in Japanese).
24. Horie, R., Yanagi, M., Ikeda, R., Yamasaki, Y., Yamariku, Y., & Takashina, T., et al. (2014). Event-related potentials caused by kawaii feeling in watching photos selected by heart beat modulation. In *Proceedings of EMBC 2014, TD18.19, Chicago*.
25. Yanagi, M., Yamasaki, Y., Yamariku, Y., Takashina, T., Hirayama, & Y., Horie, R., et al. (2014). Physiological responses caused by kawaii feeling in watching photos. In *Proceedings of AHFE 2014* (pp. 839–847).
26. Takashina, T., Yanagi, M., Yamariku, Y., Hirayama, Y., Horie, R., & Ohkura, M. (2014). Toward practical implementation of emotion driven digital camera using EEG. In *Proceedings of AH 2014* (Article No. 3).
27. Yanagi, M., Yamariku, Y., Takashina, T., Hirayama, Y., Horie, R., & Ohkura, M. (2014). Relation between kawaii feeling and heart rates in watching photos the second report-change in the heart rate by the kind of kawaii photos. In *Proceedings of the 16th Japan Society of Kansei Engineering Conference and General Meeting, E11, Tokyo* (in Japanese).
28. Pollard, G., & Ashton, R. (1982). Heart rate decrease: A comparison of feedback modalities and biofeedback with other procedures. *Biological Psychology*, *14*(3–4), 245–257.

Chapter 6
Measurements and Evaluations of Kawaii Products by Eye Tracking

Tipporn Laohakangvalvit, Ikumi Iida, Saromporn Charoenpit and Michiko Ohkura

Abstract Affective values are critical factors for manufacturing in Japan. Kawaii, which is a positive adjective that denotes such positive meanings as cute or lovable, becomes even more important as an affective value (Chap. 1). Some research has evaluated kawaii feelings by various biological signals. However, since no detailed eye tracking has been conducted yet, we employed it to identify the relationship between kawaii feelings and eye movements. We previously performed an experiment on preferences in which participants chose their favorite kawaii illustrations from six choices. However, we could not perform detailed analysis due to the complexity of the eye movements and the calibration-free data. Therefore, we improved our experiment method by randomly showing only two illustrations at a time from the six choices. From our analyzed results, we clarified the relationship between kawaii feelings and eye movements and identified two new indexes.

Keywords Affective value · Eye tracking · Kawaii feeling

6.1 Introduction

This chapter is based on a journal article [1]. Kawaii is considered as one affective value that denotes such positive connotations as cute, lovable, and charming and plays an important role in the worldwide success of many products. As described in Chap. 1, systematic analysis of kawaiiness of artificial products has been performed to clarify the attributes to design kawaii products. Furthermore, our researches have systematically studied the kawaii feelings evoked by those kawaii attributes in which the biological signals were employed including heartbeats and brain waves (Chap. 5). Eye tracking has not been scrutinized yet to study kawaii feelings, but

T. Laohakangvalvit (✉) · I. Iida · M. Ohkura
Shibaura Institute of Technology, Tokyo, Japan
e-mail: nb15505@shibaura-it.ac.jp

S. Charoenpit
Thai-Nichi Institute of Technology, Bangkok, Thailand

© Springer Nature Singapore Pte Ltd. 2019
M. Ohkura (ed.), *Kawaii Engineering*, Springer Series on Cultural Computing, https://doi.org/10.1007/978-981-13-7964-2_6

it has been widely used in various research fields, including cognitive and experimental psychology, human–computer interaction, and product development, which revealed that eye tracking can recognize human emotional states and preferences. For example, the design of such daily products as wristwatches and mobile phones was evaluated using eye tracking to explore and identify the product components that attract user attention [2, 3]. Therefore, eye tracking is an effective method for evaluating the mental states or the implicit needs of people.

We employed eye tracking to evaluate kawaii feelings to clarify the relationship between kawaii feelings and eye movements. Previously, we experimentally determined subjective preferences for kawaii illustrations by recording eye tracking while participants chose the most kawaii illustration from six on display [4] (Fig. 6.1). The result clarified the differences between favorite and most kawaii illustrations as well as the differences in preferences between genders. However, the accuracy of the eye-tracking result from a calibration-free eye-tracking device was insufficient to scrutinize and clarify the relationship between kawaii feelings and eye movements because the six illustrations were shown simultaneously, which complicated the eye movements.

To solve our previous experiment's problem, we improved our method by showing at a time only two illustrations randomly selected from the six options. We also enlarged the two illustrations to show their details more clearly. We calibrated the eyes of all of the participants before starting the illustration evaluations and tracked

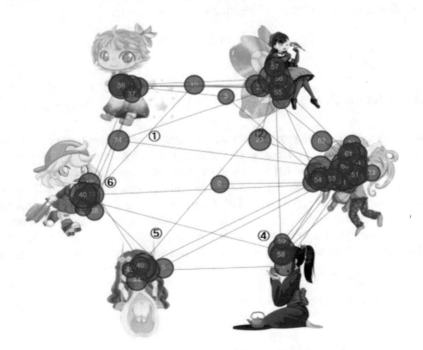

Fig. 6.1 Eye movements recorded during an evaluation of six kawaii illustrations

their eye movements while they chose the more kawaii illustration from the pair being displayed. Finally, they answered a questionnaire about their illustration choices, the most kawaii illustration, and their favorite illustration. From the experimental results, we analyzed the eye-tracking data, the cumulative kawaii scores, and the questionnaire results. This chapter describes our experiment that clarified the relationship between kawaii feelings and eye movements.

6.2 Experimental Method

6.2.1 Comparison System

A comparison system was modified from a system that evaluated kawaii ribbons [5]. As visual stimuli, that system used six illustrations (No. 1–6) (Fig. 6.2), which were original drawings to eliminate potential preference bias from famous cartoon characters.

The six illustrations were displayed in pairs with left–right counterbalanced. The total number of compared pairs was 30. All of the system content was described in Japanese. The structure of the system is described as follows:

(1) Top page: questionnaire explanation.
(2) Selection of participant's gender and age.

Fig. 6.2 Original illustrations used in comparison system to select of kawaii illustrations. Sharp sign (#) with number denotes illustration number that represents six illustrations, #1, #2, #3, #4, #5, and #6, as shown above each illustration

Fig. 6.3 Example of page in comparison system showing two illustrations and selection arrows

(3) Explanation of illustration selections: the illustrations were displayed in pairs for five seconds. Selection of more kawaii illustrations was performed using the keyboard's left or right arrow keys.

(4) Illustration selection: 30 pairs were randomly displayed for each participant. An example of this page's screenshot is shown in Fig. 6.3.

(5) Questionnaire: three subjective questions were asked: reason for selecting the illustrations (free description), most kawaii illustration, and favorite illustration.

After the participants submitted their questionnaires, the results of the illustration selections and the questionnaires were saved in a database.

6.2.2 Experimental Setup

Figure 6.4 shows the experimental setup. The questionnaire was accessed from the eye-tracking system through a web browser, i.e., Google Chrome, whose system ran on a separate PC due to limited resources. The eye-tracking system employed the EyeTech TM3 nonintrusive eye tracker (EyeTech Digital Systems, Inc.) and QG-PLUS software (DITECT Co., Ltd.) to record the eye movements and display the eye-tracking data. We used a 19-inch LCD monitor with resolution of 1280 × 1024 pixels.

6.2.3 Experimental Procedure

The following are the experimental procedures:

1. Participants sit on chairs in front of the PC.
2. They read the explanation sheet.

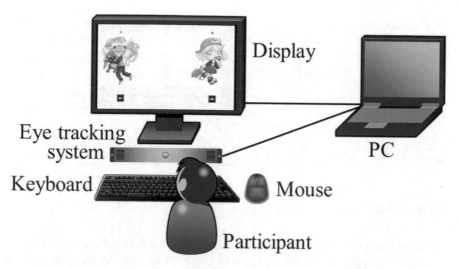

Fig. 6.4 Experimental setup where participant looks at illustrations on PC monitor while eye-tracking system records his eye movements

3. Experimenter calibrates the eyes of the participants.
4. Experimenter shows the comparison system and starts recording the eye tracking.
5. They answer their general information: gender and age.
6. They select from 30 pairs of illustrations.
7. They answer the questionnaires on three questions as follows:

   - Which is the most kawaii illustration?
   - Which is the favorite illustration?
   - What are your reasons to select kawaii illustrations? (free description).

8. Experimenter stops recording the eye tracking.

6.3 Experimental Results

6.3.1 Participants

The experiment was performed with 38 volunteers: 14 males in their 20s, 10 females in their 20s, and 14 females 65 years or older. However, only 21 bits of eye-tracking data (7 males in their 20s, 8 females in their 20s, and 6 females 65 years or older) were successfully collected due to eye calibration or eye-tracking failures.

Table 6.1 Rankings based on kawaii scores, which are cumulative values as total number of selections. Sharp sign (#) with number denotes illustration number. Number inside parentheses () show data used for ranking kawaii scores

Participant group	Ranking of illustration: kawaii scores					
	1st	2nd	3rd	4th	5th	6th
Male 20s	#3 (90)	#2 (79)	#4 (73)	#6 (63)	#5 (59)	#1 (56)
Female 20s	#2 (69)	#4 (60)	#1 (51)	#3, #6 (41)		#5 (38)
Female over 65	#1 (99)	#3 (77)	#6 (69)	#2 (67)	#4 (63)	#5 (45)

6.3.2   Cumulative Results

The cumulative results (the kawaii scores) were collected from the total number of illustration selections from 38 participants and used to rank the illustrations. The rankings, based on cumulative kawaii scores, are shown in Table 6.1. The rankings among the three participant groups were quite different.

6.3.3   Questionnaire Results

The questionnaire results consist of the following three items:

(1)   Number of illustrations selected as the more kawaii.
(2)   Number of illustrations selected as favorites.
(3)   Reasons for selecting more/favorite illustrations (free description).

The results of (1) and (2) were used to rank the illustrations, as shown in Tables 6.2 and 6.3. The ranking results showed that the first and last rankings of Tables 6.1 and 6.2 are similar. Furthermore, the first rankings for all of the participant groups of Table 6.3 were the same, which shows that all three participant groups preferred illustration #4.

Table 6.2 Ranking based on illustrations selected as more kawaii. Details are identical as described in Table 6.2, where data used for ranking are "number of illustrations selected the most kawaii" from questionnaire item 1

Participant group	Ranking of illustration: kawaii scores					
	1st	2nd	3rd	4th	5th	6th
Male 20s	#3, #4 (4)		#6 (3)	#2 (2)	#5 (1)	#1 (0)
Female 20s	#2 (4)	#3, #6 (2)		#4, #5 (1)		#1 (0)
Female over 65	#1 (4)	#3, #6 (3)		#2, #4 (2)		#5 (0)

Table 6.3 Ranking based on illustrations selected as favorites. Details are identical as described in Table 6.2, where data used for ranking are "number of illustrations selected as favorites" from questionnaire item 2

Participant group	Ranking of illustration: kawaii scores					
	1st	2nd	3rd	4th	5th	6th
Male 20s	#4 (5)	#3 (3)	#2, #5, #6 (2)			#1 (0)
Female 20s	#4 (5)	#2 (4)	#6 (1)	#1, #3, #5 (0)		
Female over 65	#4 (6)	#3 (3)	#1, #2 (2)		#6 (1)	#5 (0)

The participants also described why they selected the illustrations (3). We summarized the results based on the number of times that they mentioned each area in the illustrations. Most participants selected the illustrations based on eye size, face shape, and hairstyle. Other selection reasons mentioned the total atmosphere, colors, gestures, costumes, facial expressions, and the baby-like deformed shape of the illustrations.

6.3.4 Result of Eye-Tracking Data

Based on the rankings of the cumulative results, we recalculated and ranked only data from the 21 participants whose eye-tracking data were successfully recorded.

We employed fixation and Area of Interest (AOI). Fixation is defined as the eye state when it remains still or looks at the same spot over a period of time (threshold) that was set to 200 ms. AOI is defined as the area used to include or exclude certain segments from analysis. For this experiment's analysis, we defined two AOIs for the left-side and right-side illustrations (Fig. 6.5) and created AOIs as ellipses based on the size of illustration #4, which is the widest and tallest. Since the shape and the size of all of the other illustrations were identical, their analysis areas were balanced.

We analyzed the eye-tracking data by employing six eye movement indexes, all of which we describe in the following sections.

Fig. 6.5 Example of AOIs of two illustrations with identical shape and size showing areas included in analysis of eye-tracking data

6.3.4.1 Total AOI Duration (Sum of Durations of All Eye Positions Inside AOI)

We analyzed the total AOI duration with two factors, the participant and illustration groups. The illustration groups include the highest kawaii score, the lowest kawaii score, the most selected kawaii illustrations, and the most selected favorite illustrations. We analyzed the total AOI duration among the illustration groups for both all participants and the grouped participants. Paired t-tests identified whether a statistically significant mean difference existed between the total AOI duration among four illustration groups. The result showed a significant difference in the total AOI duration between the illustrations with the highest and lowest kawaii scores ($p < 0.05$) for females in their 20s. For the other illustration groups and participant groups, we found no significant differences in total duration. The result of the average total AOI duration is illustrated in Fig. 6.6.

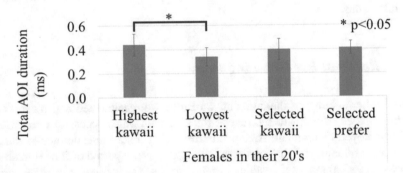

Fig. 6.6 Total AOI duration versus illustration groups of females in their 20s where highest kawaii refers to illustrations with highest kawaii scores of each participant, lowest kawaii refers to illustration with lowest kawaii scores of each participant, selected kawaii refers to illustrations selected by each participant as most kawaii, and selected prefer refers to illustrations chosen by each participant as favorites

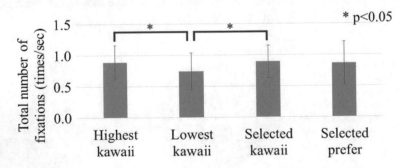

Fig. 6.7 Total number of fixations versus illustration groups of all participants (details are identical as described in Fig. 6.6)

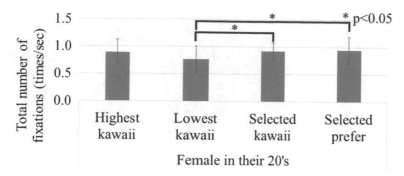

Fig. 6.8 Total number of fixations versus illustration groups and participant groups of females in their 20s (details are identical as described in Fig. 6.6)

6.3.4.2   Total Number of Fixations (Sum of All Fixations Inside AOI)

We analyzed the total number of fixations with two factors: participant and illustration groups. We performed a statistical analysis with the same method as that for the total AOI duration. The result of all participants from the paired t-tests showed a significant difference in the total number of fixations between the highest and lowest kawaii scores ($p < 0.05$) and between the most selected kawaii and lowest kawaii score ($p < 0.05$) (Fig. 6.7).

Furthermore, the result of the grouped participants from the paired t-tests showed a significant difference in the total number of fixations between the most selected kawaii and the lowest kawaii score ($p < 0.05$) and between the most favorites and the lowest kawaii score ($p < 0.05$) for females in their 20s. For other groups of illustrations and participant groups, there were no significant differences in the total number of fixations. The result of the average total number of fixations is illustrated in Fig. 6.8.

6.3.4.3   Number of Transitions Between AOIs (Sum of Times that the Eyes Moved Between AOIs for Each Pair of Illustrations)

We counted the eye movement from one position in an AOI to another position in another AOI as a transition. For an example shown in Fig. 6.9, the number of transitions is three times as the eye positions move for three times between left and right AOIs.

We analyzed the number of transitions between the two AOIs for all participants and each participant group. Since we considered the illustrations pair by pair, the difference of kawaii scores between each pair was calculated from the cumulative and questionnaire results that we used to analyze this eye-tracking metric. A Pearson product-moment correlation determined the relationship between the number of

Fig. 6.9 Example of method
to calculate number of
transitions between AOIs

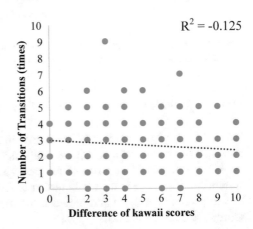

Fig. 6.10 Number of
transitions between AOIs
and differences of kawaii
scores of all participants
where red line shows
negative linear relationship
between these two variables

transitions and the differences of the kawaii scores. A scatter plot between these
two variables (Fig. 6.10) shows a linear relationship with a negative correlation. The
result shows a statistically significant ($R^2 = -0.125$, $p < 0.01$) negative correlation
between the number of transitions and the differences of kawaii scores.

We also analyzed the number of transitions between AOIs for each participant
group. The results showed similar tendencies as the result of all participants for both
males in their 20s and females 65 or older. There were negative correlations between
the number of transitions and the differences of kawaii scores with a statistical
significance for males in their 20s ($p < 0.05$) and females 65 or older ($p < 0.1$).
The result of the females in their 20s did not show a statistical difference for the
correlation. However, it did have a similar tendency as in the other participant groups,
where a negative linear relationship existed between the number of transitions and
the differences of the kawaii scores.

6.3.4.4 Number of Matchings Between Last-Eye-Position Illustrations and Selected Illustrations

We collected and analyzed the matched and unmatched numbers for each pair of
illustrations between the last-eye-position illustrations and the selected illustration
from the cumulative and questionnaire results (Fig. 6.11). Paired t-tests determined
whether a statistically significant mean difference existed between the number of

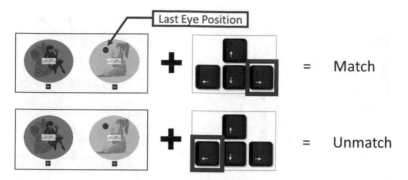

Fig. 6.11 Example of method to calculate number of matchings between last-eye-position illustrations and selected illustrations

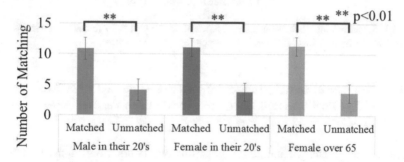

Fig. 6.12 Number of matchings between last-eye-position and selected illustration of first 15 illustration pairs for each participant

matched and unmatched selections for each participant group. The result from each participant group showed a significant difference in the number of matchings between the last-eye-position and the selected illustrations ($p < 0.01$) (Fig. 6.12).

6.3.4.5 Total Duration of Decision (Time from Occurrence of Pair of Illustrations to Selection by Arrow Keys)

We compared the total duration to identify the differences of the kawaii scores for each participant group. A Pearson product-moment correlation determined the relationship between these two variables. A scatter plot for the female participants in their 20s (Fig. 6.13) shows a statistically significant ($R^2 = -0.206$, $p < 0.05$) linear relationship with a negative correlation between the total duration of decision and the differences of the kawaii scores for females in their 20s. The results for males in their 20s and females 65 or older did not show a statistical difference for the correlation.

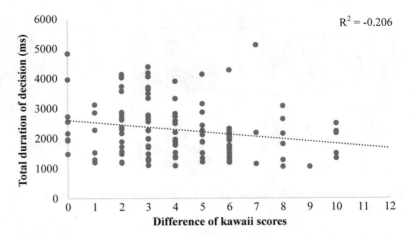

$R^2 = -0.206$

Fig. 6.13 Total duration of decision and difference of kawaii scores for females in their 20s where red line shows negative linear relationship between these two variables

6.3.4.6 Number of Initial Eye Positions on Each Focused Area (Sum of Eye Positions that Were First Inside Each Focused Area in AOI)

To collect the number of initial eye positions, we defined the focused areas that participants tended to look on their first glance. Since all of the illustrations were composed of human structure, we categorized the focused areas into three groups: head, body, and others (Fig. 6.14).

We measured and calculated the head-to-body ratios for all the illustrations using head and body heights (Fig. 6.15). We also listed the objects for other areas that included the objects that surrounded the head and body. The structure information of the six illustrations is shown in Table 6.4. The head-to-body ratio showed the

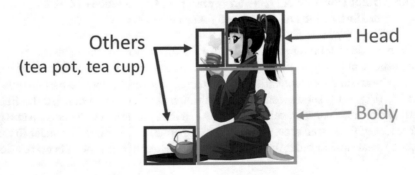

Fig. 6.14 Example of illustration divided into three focused areas (head, body, and other areas)

Fig. 6.15 Example of illustration showing method to measure head and body heights

Table 6.4 Structure information of six illustrations

Illustration no.	Head height (pixels)	Body height (pixels)	Head-to-body ratio	Surrounding objects
1	155	187	0.829	Stars
2	74	209	0.354	Wings, bird
3	88	207	0.425	Bear, pillow
4	105	244	0.430	Tea pot, tea cup
5	119	199	0.598	Rope
6	155	180	0.861	Luggage, lollipop

varieties of head and body sizes. Moreover, there were various objects in the other areas for all six illustrations.

We analyzed the number of initial eye positions using a two-factor ANOVA between the focused areas and the participant groups for each illustration. The result showed significant differences ($p < 0.01$) among the focused areas for all the illustrations. We performed a low-level analysis using the percentage of the number of first eye positions shown in Fig. 6.16. The head area tended to have the highest percentage of initial eye positions. In addition, the graph showed a tendency between genders where the numbers of first eye positions on body areas were larger in the female participants than in the male participants.

6.4 Discussion

Analysis of our rankings shows the similarity of the first and last ranking tendencies between the results of the cumulative kawaii scores and the more kawaii illustrations from the questionnaire results. In addition, the first ranking tendencies of the selected favorite illustrations were similar for all the participant groups.

Our eye-tracking analysis included six eye-tracking metrics and the following results:

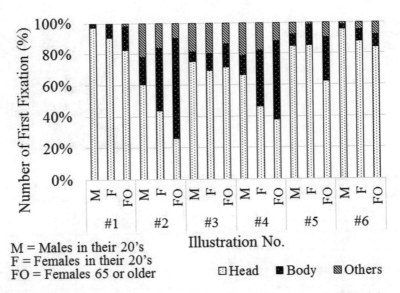

M = Males in their 20's
F = Females in their 20's
FO = Females 65 or older

⊡ Head   ■ Body   ⧅ Others

Fig. 6.16 Percentage of first fixations for all participant and illustration groups. Each bar graph shows percentage of first fixations among head, body, and other areas of each illustration

- **Total AOI duration** of females in their 20s showed significant differences between the illustrations with the highest and lowest kawaii scores. They tended to look longer at the kawaii illustrations, while the two other participant groups showed average attention to all the illustration groups.
- **Total number of fixations** of all participants showed significant differences between the illustrations with the highest kawaii scores and the selected kawaii illustrations versus illustrations with the lowest kawaii scores. The illustrations with the highest kawaii scores and the selected kawaii illustrations had more fixations.
- **Total number of fixations** of females in their 20s showed significant differences between selected kawaii illustrations and the illustrations with the lowest kawaii scores, and between their selected favorite illustrations and the illustrations with the lowest kawaii scores. They tended to look more frequently at their more kawaii selections and favorite illustrations.
- **Number of transitions between AOIs** versus the differences of kawaii scores had a significantly negative correlation for all participants, males in their 20s, and females 65 or older. For females in their 20s, the result also showed negative tendencies that resembles the other two participant groups.
- **Number of matchings between the last-eye-position illustrations and the selected illustration** showed a significant difference for the three participant groups. The larger number of matchings showed that the participants tended to take one final look at the illustrations they selected.

- **Total duration of decision** versus the differences of the kawaii scores had a significantly negative correlation for females in their 20s, who tended to take a longer time to decide if the kawaiiness of the pairs of illustrations was similar.
- **Number of first fixations for each focused area** showed a significant difference. The illustrations had various head and body sizes. All three participant groups looked more at the head areas than the other areas. Even though the head might be large or small, it attracted the most initial attention, showing that the participants looked first at the heads regardless of their sizes. The result of participants who looked first at the head area most corresponds to the result of question (3) where the participants selected illustrations based on eye size, face shape, and hairstyle. In another tendency, the females also first looked at the body and other areas in addition to the head than males, suggesting interest in other areas in addition to the head.

From our analyzed results, females in their 20s tended to look longer and more frequently at more kawaii illustrations, suggesting that they have the strongest interest in kawaii. All participants repeatedly compared the illustrations before selecting when the kawaiiness of two illustrations was similar. Finally, our results also revealed that females tend to judge kawaiiness based on the complete atmosphere while males focused on faces.

As a result, we clarified the relationship between kawaii feelings and eye movement indexes and identified two new indexes: (1) number of transitions between AOIs and (2) number of matchings between last-eye-position illustrations and selected illustrations.

6.5   Conclusion

This chapter introduces the improvement of our previous study on the study of kawaii feelings using eye tracking. We experimentally used a comparison system with an eye-tracking system and used the cumulative results, the questionnaire results, and the recorded eye-tracking data for analysis. We clarified the relationship between kawaii feelings and eye movements.

Females in their 20s tended to look longer and more frequently at more kawaii and their favorite illustrations. All participants tended to take longer to compare pairs whose kawaiiness was similar. The participants all tended to take one last look at the illustration they selected. Finally, all participants tended to focus on the head area when they first looked at the illustrations.

During this experiment, we faced some problems that might complicate collecting and analyzing eye-tracking data:

- Identical illustrations were displayed at the same position during two consecutive pairs.
- The participants did not have default eye position before they started evaluating the illustrations.

- Some participants used a mouse instead of a keyboard during the selection of illustrations which might cause slower eye movements and increase cognitive workload for the hand–eye interactions.
- Ceiling-light- and sunlight-affected devices that failed to calibrate or track eye movements.

Future work will improve our experiment based on these observations. We will collect a greater amount of eye-tracking data for further analysis between kawaii feelings and eye movements.

Acknowledgements Part of this work was supported by Grant-in-Aid Scientific Research Number 26280104. We thank José Estevão Pinto de Oliveira and Juliene Ivy Figueiredo Moreira for the original illustrations. We also thank the Koto-ku Silver Human Resources Center and the students of Shibaura Institute of Technology for their participation.

References

1. Laohakangvalvit, T., Iida, I., Charoenpit, S., & Ohkura, M. (2017). A Study of kawaii feeling using eye tracking. *International Journal of Affective Engineering, 16*(3), 183–189.
2. Köhler, M., Falk, B., & Schmitt, R. (2015). Applying eye-tracking in Kansei engineering method for design evaluations in product development. *International Journal of Affective Engineering, 14*(3), 241–251.
3. Kukkonen, S. (2005). Exploring eye tracking in design evaluation. In *Proceedings of Joining Forces—International Conference on Design Research*.
4. Ohkura, M., Iida, I., Sano, H., & Morishita, W. (2015). Comparison of preferences of kawaii characters by eye tracking between genders and generations of Japanese. In *Proceedings of the 37th Annual International Conference of the IEEE Engineering in Medicine and Biology Society*.
5. Ohkura, M., Komatsu, T., Tivatansakul, S., & Charoenpit, S. (2012). Comparison of evaluation of kawaii ribbons between genders and generation of Japanese. In *Proceedings of 2012 IEEE/SICE International Symposium on System Integration (SII)* (pp. 824–828).

Chapter 7
Meaning of "Kawaii" from a Psychological Perspective

Hiroshi Nittono

Abstract The classical explanation is that cuteness is based on feelings instinctively induced by baby schema. Contrarily to this, the Japanese concept of kawaii has a broader range of meanings in addition to cuteness and is best conceptualized as an emotion reflecting a person's perception of his or her relationship with an object. This article provides a brief description of recent trends in psychological research on kawaii and cuteness.

Keywords Cuteness · Emotion · Baby schema

You might have heard of the Japanese word *kimo-kawaii*. It means "creepy but cute," connecting words of opposite meanings. Moreover, the word *busa-kawaii* means "ugly but cute." Therefore, although the word *kawaii* is often translated into English as "cute," it appears to include a broader range of meanings in addition to cute. The author has developed a psychological model to explain the seemingly diverse connotations of the word *kawaii*, in which it is defined as an emotion [4].

The psychology of cuteness has its roots in the concept of *Kindchenschema* or baby schema, which was introduced by the Austrian ethologist Konrad Lorenz over 75 years ago [2]. As shown in Fig. 7.1, Lorenz advocated that human beings instinctively perceive physical features such as a large head compared to the size of the body, a prominent and large forehead, and big eyes positioned below the center of the face as cute, and that these features elicit caretaking behaviors. This concept was introduced as an example to describe Lorenz's theory that each animal has a critical stimulus (called a *sign stimulus*) that induces a specific fixed action pattern.

In contemporary Japan, however, the word *kawaii* is used unrelated to baby schema, for example in fashion design, among other uses. Moreover, not all

This article is a revised translation from the article written in Japanese published in M. Ohkura (Ed.)., *Kawaii Kohgaku* (*Kawaii Engineering*) published in 2017 from Asakura Shoten (Tokyo, Japan).

H. Nittono (✉)
Graduate School of Human Sciences, Osaka University, Osaka, Japan
e-mail: nittono@hus.osaka-u.ac.jp

Fig. 7.1 Visual stimulus
features that were thought to
elicit caretaking responses
[2]. Illustrations in the left
show characteristics of baby
schema that are perceived as
cute or kawaii. (Reproduced
with permission from Wiley)

people perceive the same object as kawaii. How can we explain these phenomena? The author and colleagues have investigated the relationship between
infantility and the feeling of cuteness of different objects (93 in total) and reported
a moderate positive correlation between infantility and the cuteness rating scores of
these objects. People tend to perceive infantile objects as cute. On the other hand,
there are also objects with low infantility and high cuteness, such as smiles. The
cuteness score in both genders for smiles was the highest among the 93 objects.
Moreover, the authors requested participants to imagine scenes in which they
interacted with various kawaii objects and evaluated their psychological states. The
results indicated that participants consider an object kawaii when they want to get
close to it or keep it around, and not when they want to help or take care of the
object. The motivation to approach an object is considered to be at the core of the
feeling of kawaii. It has also been reported that people tend to look at kawaii photos
for a longer time (see [4], for detailed descriptions of these studies).

When people look at something they consider to be kawaii, zygomaticus major
muscle activity increases and the corners of the mouth are lifted, which result in a
smiling face. As indicated in the survey described above, smiles are evaluated as
kawaii. Therefore, when people think that an object is kawaii, they smile, and a
smiling face, in turn, is perceived as kawaii. This exciting correlation suggests that

kawaii is an emotion that is amplified by social interactions. The author has named this phenomenon the "kawaii spiral" [4].

The findings discussed above as a whole suggest that kawaii can be regarded as a positive emotion characterized by a social approach motivation that is devoid of tense feelings or feelings of being threatened by the object. Objects that produce such an affective feeling might be comprehensively referred to as kawaii, regardless of whether they are relevant to baby schema. Differences between *kimoi* (creepy) and *kimo-kawaii* (creepy but cute) described above could be explained by this definition. *Kimoi* is merely a negative expression, whereas *kimo-kawaii* expresses the feeling that "Although other people may say it is creepy and gross, I'm interested in it and want to have a closer look at it." When the suffix *kawaii* is added to an expression, it communicates a person's positive attitude towards the object. There are no correct answers to whether an object is kawaii or not. As a result, kawaii can be used for anything in daily life for softening negative language.

Considering kawaii an emotion can explain various phenomena that cannot be explained by the theory of baby schema. Then, why has the "kawaii culture" become popular in Japan? It has been pointed out that Japanese people are characterized by *amae* (psychological dependence), i.e., behaviors and motivations for obtaining affection and acceptance from others, and *chijimi shikou* (preference for making things compact), i.e., the tendencies to feel affection for small and touchable things. The author considers that compared to other countries and cultures, the feeling of kawaii has been socially accepted and attracted attention in Japanese society because of the influence of Japanese tradition.

Recently, the number of studies on cuteness in the field of psychology and cognitive neuroscience has increased, and these studies have mainly investigated infantile cuteness (see [1], for a review). On the other hand, certain researchers are also investigating variables other than baby schema. For example, Sherman and Haidt [5] suggested that mentalizing, or the process of guessing the mental state of others, is facilitated when we look at cute things, which leads to considerate behavioral tendencies. Such changes lead to identifying others as one's peers. Moreover, Nenkov and Scott [3] conducted a study on cuteness unrelated to infantility, which to the best of the author's knowledge is the first study conducted outside of Japan from this perspective. They named unique characters and pop designs for example as "whimsical cuteness," which induces the feeling of fun and self-indulgent behaviors, such as scooping more ice cream from a container to eat and choosing rich over healthy food for dinner. According to them, cuteness relevant to baby schema, on the contrary, would result in more careful and self-restraint behaviors.

Kawaii is a concept with a broader range of meanings than cuteness. It goes beyond the response to infantile stimuli and is better conceptualized as a more general, positive emotion related to sociality and approach motivation. It is expected that studies on kawaii and cuteness based on perspectives of humanities and social sciences, as well as engineering and information science, will increase in the future because this emotion is expected to gain importance in the mature society of symbiosis.

References

1. Kringelbach, M. L., Stark, E. A., Alexander, C., Bornstein, M. H., & Stein, A. (2016). On cuteness: Unlocking the parental brain and beyond. *Trends in Cognitive Sciences, 20,* 545–558. https://doi.org/10.1016/j.tics.2016.05.003.
2. Lorenz, K. (1943). Die angeborenen Formen möglicher Erfahrung [The innate forms of potential experience]. *Zeitschrift für Tierpsychologie, 5,* 235–409. https://doi.org/10.1111/j.1439-0310. 1943.tb00655.x.
3. Nenkov, G. Y., & Scott, M. L. (2014). "So cute I could eat it up": Priming effects of cute products on indulgent consumption. *Journal of Consumer Research, 41,* 326–341. https://doi. org/10.1086/676581.
4. Nittono, H. (2016). The two-layer model of 'kawaii': A behavioral science framework for understanding kawaii and cuteness. *East Asian Journal of Popular Culture, 2,* 79–95. https://doi.org/ 10.1386/eapc.2.1.79_1.
5. Sherman, G. D., & Haidt, J. (2011). Cuteness and disgust: The humanizing and dehumanizing effects of emotion. *Emotion Review, 3,* 245–251. https://doi.org/10.1177/1754073911402396.

Part III
Applications of Kawaii Engineering

Chapter 8
Kawaii Spoons for Senior Citizens

Michiko Ohkura, Somchanok Tivatansakul and Kohei Akimoto

Abstract Senior citizens require a certain amount of daily nutrition and calories to maintain their health, especially because they often suffer from reduced appetites. Since using kawaii spoons might increase their appetites, while they looked at various kawaii spoon designs, we experimentally estimated their mental states by measuring their biological signals.

Keywords Kawaii spoon · Senior citizen · Appetite · ECG

8.1 Introduction

This chapter introduces our experiment based on the contents of our previous work [1].

As mentioned in the previous chapters in Part I, the authors focused on kawaii as an affective value of industrial products and have been conducting research to systematically analyze their physical attributes. In these researches, to subjective evaluate kawaii, we used a questionnaire with a grading scale method that resembles the subjective evaluation for other affective words such as calm and enjoyable. In addition, we used more objective biological signals and obtained research results on the relationship between kawaii feelings and biological signals (ECG, EEG).

The third author has been manufacturing and selling such tableware as spoons that are easy to use for people who have difficulty using their hands.

Elderly people in such welfare facilities as day-care centers require a certain amount of daily nutrition and calories to maintain their health, especially because they often lack appetites. Since many Japanese people tend to feel excited when they look at something kawaii, we thought using kawaii spoons might increase their appetites. Therefore, we performed an experiment during which seniors looked at

M. Ohkura (✉) · S. Tivatansakul
Shibaura Institute of Technology, Tokyo, Japan
e-mail: ohkura@sic.shibaura-it.ac.jp

K. Akimoto
Aoyoshi Co., Ltd, Tsubame, Japan

© Springer Nature Singapore Pte Ltd. 2019
M. Ohkura (ed.), *Kawaii Engineering*, Springer Series on Cultural Computing, https://doi.org/10.1007/978-981-13-7964-2_8

kawaii spoons and we estimated their mental states with such biological signals as ECG and EEG. This chapter introduces an experiment based on the contents of our previous work [1].

8.2   Experimental Method

We use four spoons from Aoyoshi Co., Ltd. for the experiment:

(1) Plain (P),
(2) Glitter gray (A),
(3) With a red ruby (B), and
(4) With a pink ribbon (C).

Three spoons (A, B, and C) were decorated to be kawaii (Fig. 8.1).
The following are the steps of our experimental procedure:

(1) We explained the experiment and obtained consent from our participants.
(2) Each participant sat at a table and was told to imagine eating rice porridge using a spoon. We explained that we would show four different spoons individually.
(3) We attached sensors to the participants, measured their ECG and EEG signals, and confirmed that the sensors were functioning properly.
(4) We began the measurements.
(5) After 40 s, we placed spoon (P) on the table next to a bowl and asked the participant to look at the spoon.
(6) After 20 s, we placed spoon (A) on the table next to spoon (P) and asked the participant to look at it.
(7) After another 20 s, we placed spoon (B) next to spoon (A) and asked the participant to look at it.
(8) After another 20 s, we placed spoon (C) next to spoon (B) and asked the participant to look at it.
(9) After another 20 s, we asked the participant about her impressions of each spoon as well as her favorite one and the reason.

Fig. 8.1   Decorated spoons

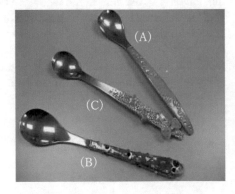

We counterbalanced the order at which the spoons were placed on the table and measured the ECGs and EEGs for around three minutes from steps (4) to (8). We used an RF-ECG EK (sampling rate of 100 Hz, and input impedance exceeds 100 MOhm) of GM3 CO., Ltd. as an ECG sensor.

8.3   Experimental Results and Discussion

We performed our experiment at a day-care center in Niigata prefecture in January 17th, 2013 with six female participants whose ages ranged from 75 to 80. All received explanations of the experiment beforehand and came to the center to participate in it.

Initially, all of the participants claimed that spoon (A) was their favorite. However, during free discussions after the experiment, some of the women admitted that they actually thought that spoons (B) or (C) were their favorites, but they deemed such choices as inappropriate to their age. So, we asked them again to honestly tell us which spoon was their favorite.

We omit the explanation of the EEG results because they were not clear. For the ECGs, we averaged the heart rates (HRs) while they observed each spoon for 20 s. Table 8.1 shows the results and Fig. 8.2 shows the differences in the averaged HRs between each decorated spoon and the plain spoon (P).

We obtained the following results:

- The averaged HRs were different among the spoons for all the participants.
- For participants 1–3, while they observed spoons (A), (B), and (C), the averaged HRs were higher than when they looked at spoon (P). For participants 4–6, the highest averaged HRs were found while they looked at spoons (B) or (C).
- No participants showed the highest averaged HR while they looked at spoon (A).
- For participants 2 and 4, their favorite spoon was (C) and they showed the highest averaged HRs while they looked at it.

From Scheffe's method of paired comparison, we obtained the following significant differences at 1 or 5% levels:

- between spoon (C) and the other spoons for participant 2,
- between spoons (B) and (P) for participant 3, and

Table 8.1 Averaged heart rate (bpm)

Participant	(P)	(A)	(B)	(C)
1	77.5	78.9	80.1	78.0
2	70.6	73.0	72.3	80.4
3	75.0	75.9	76.4	75.6
4	79.1	78.2	77.6	79.6
5	96.9	94.4	98.1	94.6
6	75.4	73.6	75.8	75.5

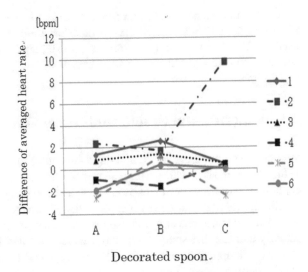

Fig. 8.2 Differences between averaged HRs while observing decorated and plain spoons

- between spoon (A) and the other spoons for participant 6.

As we have already described in the previous chapter (Chap. 3), HR increases not only by experiencing anxiety and stress but also by feeling excited about kawaii objects. The experimental results in this chapter can be interpreted as the relation of the HR increase and feeling excited while looking at kawaii spoons. We estimated their favorite spoons from ECGs based on the increases of their averaged heart rates. The finding of this experiment is valuable because it verifies that kawaii is an affective value for both the young and the old.

8.4 Conclusion

From the experimental results for six elderly women, we estimated their favorite spoons from ECGs by the increases of their averaged heart rates.

Because our experiment featured different-shaped spoons and an insufficient number of participants, future work will perform another experiment under a more controlled environment.

Acknowledgements We thank the visitors of Tsubame Welfare Volunteer Center who have volunteered to participate in the experiment.

Reference

1. Ohkura, M., Tivatansakul, S., & Akimoto, K. (2014). Kawaii spoons and heart rates of elderly. *The IEICE Transactions on Information and Systems, 97*(1), 177–180. (in Japanese).

Chapter 9
Toward Practical Implementation of an Emotion-Driven Digital Camera Using EEG

Tomomi Takashina, Miyuki Yanagi, Yoshiyuki Yamariku, Yoshikazu Hirayama, Ryota Horie and Michiko Ohkura

Abstract Photography is closely tied to the emotions of individuals. Therefore, an emotion-driven camera can be a natural consequence of photography. In the field of electroencephalogram (EEG) emotion detection in research laboratories, there have been many works; however, such methodologies are considered to be difficult to apply in a real-world environment. One of the primary issues is the cue problem. In the real world, it is difficult to detect cues in a general application. Our proposed design enables the detection of cues for digital cameras. The proposed equipment has a viewfinder that has the capability of observing the eye using a proximity sensor, an image-processing module to detect a scene change, and an EEG headset. This system enables the reproduction of an environment similar to that of the research laboratory while still being a natural configuration for an ordinary digital camera.

Keywords Affective computing · Brain–computer interface · Camera · Kawaii feeling

9.1 Introduction

This chapter is mainly based on the reference [1]. Photography is closely tied to the emotions of individuals. Therefore, an emotion-driven digital camera can be a natural consequence of photography. Such a camera would record an image of the scenery when a user experiences certain emotions.

The key element of the emotion-driven digital camera is how to detect emotions from vital reactions. Yanagi et al. investigated the detection of an emotion using electrocardiography (ECG) of participants while they are looking at photographs (Sect. 5.6) [2]. However, ECG has limitations if we want to use it to detect wide

T. Takashina · Y. Yamariku · Y. Hirayama
Nikon Corporation, Yokohama, Japan

M. Yanagi · R. Horie · M. Ohkura (✉)
Shibaura Institute of Technology, Tokyo, Japan
e-mail: ohkura@sic.shibaura-it.ac.jp

© Springer Nature Singapore Pte Ltd. 2019
M. Ohkura (ed.), *Kawaii Engineering*, Springer Series on Cultural
Computing, https://doi.org/10.1007/978-981-13-7964-2_9

Fig. 9.1 EEG measurement in laboratory

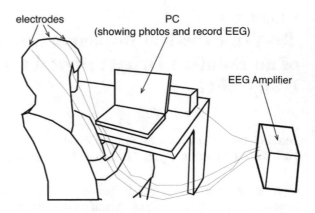

varieties of emotions. Therefore, an electroencephalogram (EEG) is more suitable because it has the potential to yield rich information.

Several prior studies in the field of EEG emotion detection in research laboratories exist [3, 4]. However, applying such methodologies in a non-laboratory environment is considered to be difficult. One of the primary issues is the cue problem. In the typical experimental environment at a research laboratory, a visual stimulus is presented to a participant at time t_s and we can use t_s as the starting point of a time series of EEG for detecting emotions. Thus, we naturally know cues in such environments.

In the real world, it is difficult to know such cues in general applications, but our proposal enables us to detect cues in the operation of a digital camera. In this proposal, the equipment has a viewfinder to observe the eye with a proximity sensor, an image-processing module to detect scene changes, and an EEG headset. This system enables us to reproduce an environment similar to that of the research laboratory while still having a natural configuration of an ordinary digital camera.

9.2 Emotion Detection by EEG

To test the feasibility of emotion detection by EEG, we measured the EEG of a participant when kawaii and uninteresting samples are presented. Kawaii is a Japanese word for an emotional value that has positive meanings, such as cute, lovable, and charming. Kawaii is one of the typical emotions that people feel while taking photos.

Figure 9.1 shows the typical configuration for EEG measurement in a laboratory. The participant looks at a photograph on a personal computer (PC) display. He or she is instructed to gaze at the center of the display to avoid the effects of electrooculography (EOG) on the EEG due to eye movement. The electrodes from the EEG amplifier are attached to the scalp of the participant. The PC records the EEG of the participant and the starting time at which each photograph is shown during the experiment.

Fig. 9.2 EEG for kawaii and
uninteresting stimuli

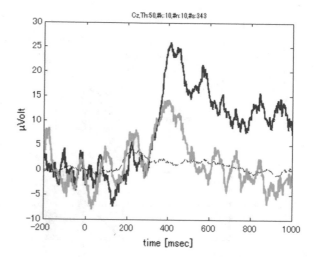

Fig. 9.2 EEG for kawaii and uninteresting stimuli

Figure 9.2 shows a typical example of EEG for a kawaii and an uninteresting photo. Blue and green lines represent the EEG for kawaii samples and the EEG for uninteresting samples, respectively. The X-axis is time in milliseconds (*ms*) and the Y-axis is signal intensity in microvolts (μV). *Time* = 0 represents the instant at which a stimulus is presented. From Fig. 9.2, there is a possibility to distinguish the feeling between kawaii and uninteresting using EEG [5]. However, as signal processing itself is not the focus of this chapter, we will not deal further with the issue.

9.3   System Design

Based on the assumption that we can detect specific emotions using EEG, we propose a system design of an emotion-driven single-lens reflex digital camera, as shown in Fig. 9.3. The point of this design is how to detect the time that a subject appears or changes on the screen and the user actually gazes the subject.

In this design, the sub-image sensor obtains an image through a prism and finds an appearance or a change of the subject, and the proximity sensor detects when the user puts his or her eyes to the viewfinder. If both conditions are true, we can identify the event as the cue of visual stimuli. Then, the emotion detector distinguishes specific feelings using the EEG recorded for a defined duration starting from the time of the cue. If any specific feeling is recognized, it sends a signal to the camera controller.

The structure of the single-lens reflex camera is suitable for use as an emotion-driven digital camera for the following reasons:

1. The viewfinder creates an environment that allows the user to concentrate on a subject. It is important because an EEG is easily disturbed by surrounding noisy factors.

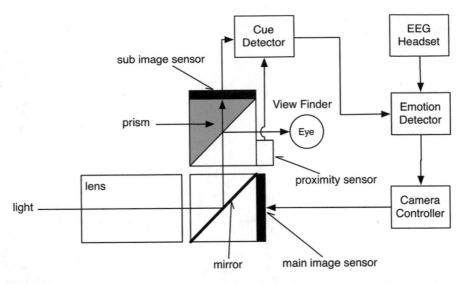

Fig. 9.3 System design of emotion-driven digital camera

2. When the user puts his or her eyes to the viewfinder we can detect that the user is actually gazing at images through the finder.
3. It is possible to have the EEG sensors built into or attached to the camera. Electrodes attached to the camera would be designed to make contact with the forehead.
4. A single-lens reflex camera generally has a sub-image sensor for exposure metering, which can be used to detect subject appearance and change.

9.4 Preliminary Prototype

We constructed a preliminary prototype consisting of an Emotiv EPOC headset, a Nikon D7000 camera, and a PC. As a preliminary research, we use EPOC's Cognitv suite to detect the user's motor imagery (i.e., thinking to push something) instead of emotion, and linked it to a camera control program. The current prototype does not yet include cue detection and emotion detection mechanisms. When the EPOC detects the user's intention to push, the PC sends a shutter release command to the digital camera, and then the digital camera shoots a photo of the subject (Fig. 9.4).

Fig. 9.4 Preliminary prototype consisting of an Emotiv EPOC headset, a Nikon D7000 camera, and EPOC's Cognitv software suite and a PC

9.5   Future Work

In future work, we will develop the emotion detection algorithm and will implement the cue detector. Though we only showed the EEG example for kawaii as an emotion in this chapter, we will extend the target emotions to the other generic emotions.

9.6   Conclusion

We proposed a realistic implementation of emotion-driven digital camera. This system enables us to reproduce an environment similar to the research laboratory while still having a natural configuration for an ordinary digital camera. In future work, we will construct the system based on the design described in this chapter.

References

1. Takashina, T., Yanagi, M., Yamariku, Y., Hirayama, Y., Horie, R., & Ohkura, M. (2014). Toward practical implementation of emotion driven digital camera using EEG. In *Proceedings of the 5th Augmented human International Conference,* AH'14, No. 31, ACM (New York, NY, USA).
2. Negishi, Y., Dou, Z., & Mitsukura, Y. (2012). EEG feature extraction for quantification of human-interest. In *Proceedings of 2012 RISP International Workshop on Nonlinear Circuits, Communications and Signal Processing (NCSP'12).*
3. Yanagi, M., Yamariku, Y., Takashina, T., Hirayama, Y., Horie, R., & Ohkura, M. (2014). Physiological responses caused by kawaii feeling in watching photos. In *Proceedings of the 5th International Conference on Applied Human Factors and Ergonomics,* pp. 839–848.
4. Yanagi, M., Yamazaki, Y., Yamariku, Y., Takashina, T., Hirayama, Y., Horie, R., & Ohkura, M. (2013). Relation between kawaii feeling in watching pictures. In *Proceedings of the 15th Annual Conference of Japan Society of Kansei Engineering.* (in Japanese).
5. Mikhail, M., El-Ayat, K., El Kaliouby, R., Coan, J., & Allen, J. J. B. (2010). Emotion detection using noisy EEG data. In *Proceedings of the 1st Augmented Human International Conference,* AH'10, No. 7, ACM (New York, NY, USA).

Chapter 10
Evaluation of Key Emotional Value for Saudi Women

Enayyah Barnawi and Michiko Ohkura

Abstract In recent years, Kansei engineering has become crucial in industrial fields. It works as a new value axis that differs from such conventional ones as functionality and price that have served as competitiveness sources in manufacturing. Based on the vast growth in the apparel industry and greater customer awareness and intelligent observation of various products, such a conventional value axis as functionality and price is insufficient to satisfy customer's needs. It is now crucial to study consumers' Kansei or emotional values and build it within products. In this chapter, we evaluated the key emotional values that influence Saudi women to purchase luxury fashion brands. In addition, we aim to define kawaii from Saudi women's perspective and identify the desired key emotional values for Saudi women and their importance compared to kawaii value. We used a qualitative method approach to gather a comprehensive understanding of the subjects of our study. A question framework was designed and in-depth interviews were employed to collect data.

Keywords Emotional value · Saudi females · Luxurious fashion brand · Society · Kawaii · Social perception

10.1 Introduction

This chapter introduces the research results described in [1] and [2]. We evaluated the key emotional values that influence Saudi women to purchase luxury fashion brands [1]. In addition, we aim to define kawaii from Saudi women's perspective and identify the desired key emotional values for Saudi women and their importance compared to kawaii value [2].

E. Barnawi · M. Ohkura (✉)
Shibaura Institute of Technology, Tokyo, Japan
e-mail: ohkura@sic.shibaura-it.ac.jp

© Springer Nature Singapore Pte Ltd. 2019
M. Ohkura (ed.), *Kawaii Engineering*, Springer Series on Cultural Computing, https://doi.org/10.1007/978-981-13-7964-2_10

In recent years, Kansei engineering has become crucial in industrial fields. The Japanese Ministry of Economy, Trade and Industry (METI) believes that a special type of positive economic value is created when industrial production is derived from Kansei [3], which METI introduced as a new value axis that differs from such conventional ones as functionality and price that provide competitiveness in manufacturing [4]. This new value axis is user/consumer Kansei and probably includes the following feelings: enjoyment, sense of security, coolness, user-friendliness, or empathy with manufacturer [4].

Recently, there has been steady growth in the apparel industry. According to a factsheet from the Clean Clothes Campaign (CCCs), in 2014 worldwide, the women's clothing industry was worth 621 billion USD (497 billion EUR) [5]. Product design is also becoming more comprehensive because the customers are becoming more intelligent and market choices are becoming too many [6]. Therefore, affective production has become more complicated, and competing manufacturers are facing difficulty attracting a maximum number of customers. Thus, since product functions or prices are insufficient to attract customers, the Kansei or emotional values of consumers must be incorporated within products.

In the first research in this chapter, we identify the key emotional values that drive Saudi women to purchase certain fashion products by performing in-depth interviews with Saudi women living in Japan, collected data, and analyzed them. This study's main objective is to check the validity of our hypotheses to find answers to our research questions. In our hypotheses, we assumed that elegance, prestige, or chic are candidates for the key emotional values that motivate Saudi women to purchase fashion products. Also, we assumed that social perception greatly influences how they dress.

In the second research in this chapter, we evaluated the desired key emotional values for Saudi women when purchasing fashion brands. In addition, we defined kawaii value from Saudi women's perspective. kawaii feeling is one important Kansei value in Japan. The term kawaii has been used in Europe and the rest of the world since the 1990s when such items of Japanese popular culture as manga, extravagant street fashions, and video games began to be exported [7]. Nowadays, Japanese cartoons, animations, and drama have a significant increased presence among Saudi Arabian youth, which makes the term kawaii a familiar expression among Saudi youth, especially those who are interested in Japanese animations. Thus, it is important to understand how kawaii is observed by people from countries that are distant from Japan and have different cultural and social standards such as Saudi Arabia. As our research result, we identified the factors that influence the way Saudi women adapt different emotional values. Using a qualitative research approach, we performed in-depth interviews with Saudi women living in Japan, collected data, and analyzed it using the Grounded Theory Approach (GTA).

10.2 Evaluation of Key Emotional Value that Influences Saudi Women to Purchase Luxury Fashion Brands

10.2.1 Background

In 2015, a study was conducted to determine what motivates Saudi women to shop at Western-style shopping centers and reached the following results:

(1) the lure of expensive brand images;
(2) improved social position within society;
(3) liberalization of women's culture; and
(4) portraying themselves as open-minded [8].

The results also reflected how Saudi women believed that purchasing Western-style goods helped them maintain their social prestige.

After uniqueness and quality, the result of another study showed that emotional value was the third factor that influenced Saudi women to purchase luxury brands [9]. The participants used a set of expressions to give their desired emotional values, including luxury, prestige, status, fashionable, and elegance.

The results represented in these works can be linked to Hofstede's 1980 cultural dimension analysis of the Arab world. He classified Arab countries as having a high power distance, high uncertainty avoidance, a collectivist culture, and a masculine culture [10].

Hofstede [10] defined individualism versus collectivism cultural orientation as the "degree to which people in a country prefer to act as individuals rather than as members of groups." In a collective society, loyalty is paramount and overrides most other societal rules and regulations, and offenses cause shame and loss of face [11]. This cultural dimension reflects the importance of society perception for Saudi people and it might explain why Saudi women [8] desire to purchase expensive luxury brands just to maintain their social "position" and "prestige."

On the other hand, Hofstede defined masculinity as the "degree to which values like assertiveness, performance, success and competition, which in nearly all societies are associated with the role of men, prevail over values like the quality of life, maintaining warm personal relationships, service, care for the weak, and solidarity, which in nearly all societies are more associated with the role of women" [10]. Since Saudi is a masculine society, the load on women is not heavy since it is not their duty to provide decent life for the family. Thus, women tend to feel more relaxed, kind, and try to maintain their social status by being chic, having prestige, and feeling elegant [8, 9].

However, since Hofstede's analysis generalized Arab countries and was not specific to Saudi Arabia, a Saudi researcher investigated Saudi Arabia's cultural factors using the Hofstede cultural value dimensions and confirmed that its cultural characteristics resembled Hofstede's 1980 analysis for the Arab world [12].

Based on these facts and results, we assumed that elegance, prestige, or chic are candidates for the key emotional values that motivate Saudi women to purchase fashion products. Also, we assumed that social perception greatly influences how they dress.

10.2.2 Methodology

To deepen our understanding of our research questions, we used a qualitative method. The qualitative method, represented by interviews or case studies, aims to gather a comprehensive understanding of an individual user from multiple aspects on the basis of a limited number of case studies [13]. In this research, we used semi-structured interviews because they elicit answers close to users' real intentions by flexibly changing questions and topics based on their reactions as needed [14].

10.2.2.1 Hypothesis

In this study, we made the following two hypotheses:

H1: Elegance, prestige, or chic are candidates for the key emotional values that drive Saudi women to purchase fashion products.
H2: Social perception greatly influences how Saudi women dress.

10.2.2.2 Participants and Interview Circumstances

The participants of our study were four undergraduate Saudi women in their twenties living in Tokyo who are studying at different Japanese universities (scholarship students). All have been living in Japan for over 2 years and half.

We performed two in-depth interviews, where two participants were interviewed together as a focused group. The time interview took 1 h and 25 min, and by the second interview took 53 min. We recorded them using a voice recorder after obtaining their permission. The interviews covered different themes including kawaii, but in this chapter, we only analyzed the themes related to the participants' desired emotional values.

10.2.2.3 Interview Questions

Our interview questions checked the validity of our hypothesis and obtained supporting information about emotional values, such as what is the desired emotional value and why? What are its attributes and how did living in Japan influence it? Also, the influence of price and society on the desire to possess clothing with the desired

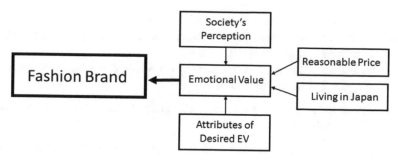

Fig. 10.1 Question framework

emotional value. Figure 10.1 illustrates the framework of our questions. We conducted semi-structured interviews, asked prepared questions, and asked follow-ups based on the participant's answers. After the interviews, the recorded data were transcribed and analyzed.

10.2.2.4   Analysis

Since our main objective at this stage is to check the validity of our hypotheses to answer our research questions, we systematically analyzed our data using qualitative content analysis [15]. To analyze the data, we performed the following procedures:

(1) transcribed the recorded voices into written texts;
(2) translated the transcribed texts from Arabic to English;
(3) performed open coding;
(4) did axial coding; and
(5) did selective coding.

We read the transcripts several times to start the open coding and mainly used a deductive approach. Then, we performed the axial coding and organized our coded data into themes, categories, and concepts. The categories were generated by the framework of the designed questions. Finally, we did selective coding by examining the coded data and answering our research questions.

10.2.3   Findings

10.2.3.1   Answers to Research Questions

We analyzed the transcripts and found the following:

1. The participants stated that they want their clothes to reflect something about themselves and argued that clothing reflects personality.

2. Chic, elegance, and luxurious are the desired emotional values that compel them to purchase luxury brands. They strongly felt that looking chic, elegant, and luxurious increases their self-confidence.
3. The participants strongly felt that almost all Saudi women care about feeling chic, elegant, and luxurious. One said:

> I think that all Saudi women care about being chic and prestigious, maybe more than any other women I've met of other nationalities.

The participants provided several reasons to explain why Saudi women care about these values:

a. The participants think that Saudi Arabia is a judgmental society, thus women choose to look chic, elegant, and prestigious to increase their self-confidence to satisfy society's perception.
b. The participants believed it might also be psychological because Saudi women usually wear abaya (traditional black female garb) in public, and the only chance for them to be chic and prestigious is in such women-only communities as parties or special occasions. This explains why they deeply care about feeling chic, elegant, and luxurious when dressing up. In the same context, one participant stated that:

> Women in our society might feel insecure and left out when they see images of fashion models in social media. So, they exploit the only chance they have to dress up in women-only societies and be chic and prestigious to compensate for feeling left behind.

Concerning why Saudi women living abroad embrace such values, they strongly felt that because the typical image of Saudi women is wearing abaya and many people know practically nothing about Saudi women, it is their duty to dress beautifully and show the world that Islam is a tolerant religion that allows them to wear beautiful clothes. However, since abroad they feel more relaxed in how they dress, they care less about their desired values because the societal pressure on them is eased.

4. Participants felt that fabric type, dress style, and color are the combination of attributes that determine whether a piece of clothing is chic, elegant, or luxurious. However, color was mentioned frequently in a different context as a positively desired attribute.
5. The prevalent feeling was that even though the desired emotional values are critical compared to price, they always try to find their desired emotional value at a reasonable price.
6. Participants felt that living in Japan improved their fashion sense due to the following: (1) influence of social media and watching how Japanese girls dress; (2) the color of clothing in Japan is more vibrant than in Saudi. One participant said:

> What I like here is the changing seasons and how colors reflect the season, and these frequent changes influenced my way of selecting colors, unlike at home.

For the effect of living in Japan on the desired emotional value, they felt that although living in Japan improved their clothing taste, it also made them more relaxed toward their desired emotional values because they feel liberated from Saudi society's perception.

10.2.3.2   Hypothesis Validity

We analyzed our transcripts and proved the validity of our hypotheses:

H1:   Elegance, prestigious, or chic are candidates for the key emotional values that drive Saudi women to purchase fashion products.

H2:   Social perception greatly influences how Saudi women dress.

Thus, **H1** and **H2** are both supported.

10.2.4   Discussions

The pie chart in Fig. 10.2 illustrates some significant frequently used words that were used positively in our interviews. The main desired emotional values that influence Saudi women to purchase fashion brands are chic, elegance, luxurious, and prestigious. Also, "self-confidence" was used nine times by the participants, which indicate its importance for them. They believed that feeling chic increases their self-confidence and their courage to face a judgmental society, and the results showed the huge influence that society's perception has on how Saudi women dress. In the context of attributes, color was mentioned 23 times, but style was mentioned only 11 times. This reflects color's importance for chic clothing since color variation was another main reason that helped to improve their taste in clothing in Japan.

Even though emotional values are critical to the participants who place it over price, they always shop for reasonable prices as much as possible. Nevertheless, participants feel more relaxed about their desired emotional values when they are abroad because they are experiencing less societal pressure. They also believe that being abroad is a good chance to reflect a positive image of their religion and country by dressing beautifully.

10.2.5   Summary

In this research, we evaluated the key emotional values that influence Saudi women to purchase luxurious fashion brands through in-depth interviews and data analysis. Our results show that chic, elegance, and luxury are the main emotional values that compel Saudi women to purchase luxury brands. Society's perception hugely influences how Saudi women dress and their desired emotional values as well. These

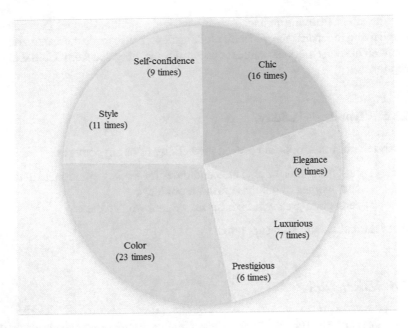

Fig. 10.2 Significant frequently used words

emotional values help to increase their self-confidence because they believe that it helps them face a judgmental society. As an important attribute, color showed a trend of determining the elegance and the luxury of clothing and was also one of the main reasons that improved the subject's taste in selecting clothing after coming to Japan.

Since the number of participants in this study was limited, it should be increased to get more generalizations for the findings. The results of this study can effectively contribute to improve the affective production in fashion industries for Saudi Arabia or other manufacturers around the world that are involved in business with Saudi Arabia. In future work, we plan to analyze our data using the Grounded Theory Approach to deepen our understanding and examine potential emerging trends from the transcripts. We will also perform the same interviews with Japanese women and compare the analyzed results to investigate the impact of cultural differences on Saudi and Japanese women.

10.3 Key Emotional Value and Kawaii for Saudi Women

10.3.1 Background

In a previous section, we represented our attempt to identify the key emotional values that drive Saudi women to purchase certain fashion products. However, the results raised an important research questions about the importance of kawaii value for Saudi women. Therefore, to investigate on this question, a group of Saudi females in Saudi Arabia were selected randomly and asked about their desired emotional values. Their answers varied between different values such as chic, elegant, prestigious, etc. but none of them was kawaii. In addition, results of two studies were found valuable for the process of this research. In 2015, a study was conducted to determine what motivates Saudi women to shop at Western-style shopping centers; the results reflected that Saudi females believed that purchasing Western-style goods helped them maintain their social prestige [8]. After uniqueness and quality, the result of another study showed that emotional value was the third factor that influenced Saudi women to purchase luxury brands [9]. The participants used a set of expressions to give their desired emotional values, including luxury, prestige, status, fashionable, and elegance.

The results from our investigation indicated that kawaii is not a critical value for Saudi females like it is for Japanese females; however, other values such chic, elegant, prestigious, etc. might be. This observation was a motivation to conduct this research.

10.3.2   Method

To satisfy the exploratory nature of this research and to get a deep understanding of research questions, a qualitative research method was used [16]. We used the same data of our semi-structured in-depth interviews which is used for the previous section (Sect. 10.2.).

10.3.2.1   Participants and Interview Questions

We rewrite this information even they are the same as the previous section. The participants were four undergraduate Saudi women in their 20s living in Tokyo who are studying at different Japanese universities (scholarship students). All have been living in Japan for over 2 years and half, which means that they have sufficient knowledge about Japanese culture including kawaii. We performed two semi-structured in-depth interviews, where two subjects were interviewed together as a focused group. The interviews were recorded using a voice recorder. The time interview took 1 h and 25 min, and by the second interview took 53 min. The interview questions had two different themes, desired key emotional values and kawaii value. The participants were asked about different details regarding these two themes.

10.3.2.2   Analysis

We analyzed the data using GTA which enables the researchers to examine different topics and behaviors from many different angles, which helps in developing comprehensive explanations. It also can be used to get more information about old problems as well as to study new and emerging areas that are needed to be investigated [16]. As an assisting tool to manage and organize the data, we used MAXQDA software. To analyze the data, we performed the following procedures:

1. transcribe the recorded voices into written text;
2. translate the transcribed texts from Arabic to English;
3. perform open coding which is an open process of giving proper concept to every piece of data; and
4. perform coding for concept development and elaboration where we refined the concepts that were defined in the open coding stage and elaborated more on them.

During the whole process, we mainly used memo writing and diagraming to develop the concepts in term of their properties and dimensions. We also performed coding for context to maintain the sensitivity of the action interactions in different contexts.

10.3.3   Results

10.3.3.1   The Key Emotional Values for Saudi Women

We sorted the sub-concepts around the core concept and we defined the relationships between them. Then, we made the diagram to illustrate the structure of the way Saudi women adapt their desired emotional values (Fig. 10.3).

The "desired emotional values" for Saudi women are chic, elegant, and luxurious. There are certain attributes that would make clothing chic, elegant, and luxurious such harmonious color (basically dark colors), fancy fabric type (such satin or silk), and certain style (for example, one or multi-pieced clothing). These attributes should have a "good quality" and come with a "reasonable price." These values are very important for Saudi women and highly demanded "in Saudi" society, which put the women in "tense" toward these values because of the fear of the "social perception." They care about these values in special occasions (women-only community) the most. The idea of women-only community exists because of the nature of Saudi society where (1) women and men have separate environments and (2) women wear abaya (traditional female garb) in public places. Therefore, the only chance for women to dress up freely (without the need of covering their hair or wearing abaya) is in women-only communities as we have already described in   Sect. 10.2. Thus, they exploit this chance and try their best to be chic and elegant which is considered as a high-class value in Saudi. However, by being abroad such in this case "in Japan," the women tend to feel "relaxed" and care less because of the absence of social pressure. In general, they believe that embracing these values would "satisfy the social perception;" it would increase their "self-confidence," which in return will provide them with more "courage to face the society."

10.3.3.2   Kawaii Value from Saudi Women's Perspective

Figure 10.4 shows the general form of the way Saudi women adapt kawaii value.

Their adaptation for kawaii value is derived by the Saudi Arabian's social standards or in other words, by what the society might say about them "social perception." These standards are set by the Saudi Arabian "cultural background." For example, women in this study recognized kawaii as a value carrying childish meaning, while in Saudi culture it is unacceptable when adults try to dress or act in a childish way, thus they manage not to buy very noticeable kawaii goods in order not to get criticized by the society. According to Saudi women, there are two "types of kawaii value:" "positive kawaii" and "negative kawaii." "Positive kawaii" can be achieved by "age consideration" (meaning kawaii is for young girls, thus adults should adapt it with limitations) and "balanced adaptation" (meaning without overdoing or overdressing). Dolly style is considered as "negative kawaii" because the lack of "balanced adaptation." The "positive kawaii" is the desired kawaii value for Saudi women that recalls the feelings of peace, relax, and simplicity. Consequently, "the desired kawaii

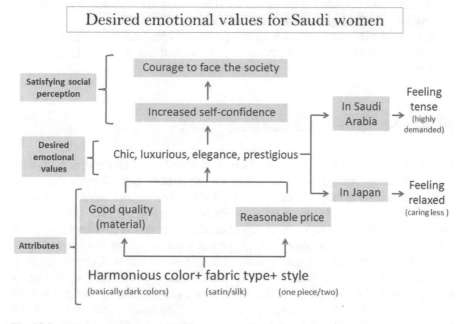

Fig. 10.3 The structure of how Saudi women adapt their desired emotional values

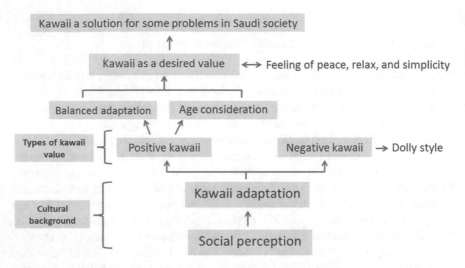

Fig. 10.4 The structure of how Saudi women adapt kawaii value in general

value" can then be utilized as a solution for some problems in Saudi society such as the problem of young girls acting ahead of their ages.

The results also confirmed that kawaii is not a critical value for Saudi women; however, chic, elegance, and luxury are crucial desired emotional values for Saudi women.

10.3.4 Discussions and Summary

In this study, we evaluated the way Saudi women adapt kawaii and their desired emotional values through in-depth interviews and data analysis. Saudi women tied kawaii value with feeling "relaxed" and their desired emotional value with feeling "tense." Feeling relaxed is a key point for Saudi women, trying to be chic, elegant, and luxurious put them in tense all the time because of the fear of social perception. In contrast, kawaii value can cause and bring peace. It gave Saudi women the feeling of peace and made them believe that importing kawaii value to Saudi would contribute in solving some problems in Saudi society. If kawaii is adapted as a culture, the feeling of peace that is evoked by kawaii should reflect on one's mentality and personality, which would reduce the violence and negativity. The results showed that social perception has significant impact on the way Saudi women adapt kawaii and their desired emotional value. They really care about what other women in their society

Fig. 10.5 Emotional value
adaptation hierarchy

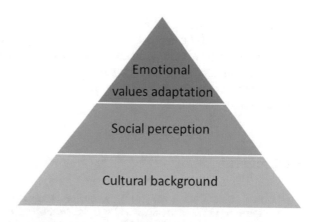

might say about them. They are willing to be chic and elegant at any expense just for the sake of increasing their self-confidence, which in return will give them more courage to face the judging society. They strictly maintain the way they adapt kawaii value. Their adaptation for kawaii is limited to what is acceptable in their culture in order not to get criticized by the society. The pyramid in Fig. 10.5 illustrates the hierarchy of how Saudi women adapt different emotional values. The cultural background works as a baseline that sets and controls society's standards, which forms the way individuals perceive each other. Thus, these individuals tend to adapt different emotional values in a way that fits the social standards and satisfy the social perception.

However, the authors believe that the impact of social perception might not last for the coming generations. Nowadays, Saudi society is exposed to globalization, and the traditional way of thinking is gradually decreasing.

10.4 Conclusions

In this chapter, we evaluated the key emotional values that influence Saudi women to purchase luxury fashion brands. In addition, we defined kawaii from Saudi women's perspective and identified the desired key emotional values for Saudi women and their importance compared to kawaii value. We used a qualitative method approach to gather a comprehensive understanding of the subjects of our study. A question framework was designed and in-depth interviews were employed to collect data. Our results show that chic, elegance, and luxury are the main emotional values that compel Saudi women to purchase luxury brands.

In addition, we evaluated the way Saudi women adapt kawaii and their desired emotional values through in-depth interviews and data analysis. Saudi women tied kawaii value with feeling "relaxed" and their desired emotional value with feeling "tense." Feeling relaxed is a key point for Saudi women, trying to be chic, elegant, and luxurious put them in tense all the time because of the fear of social perception. In contrast, kawaii value can cause and bring peace. It gave Saudi women the feeling of peace and made them believe that importing kawaii value to Saudi would contribute in solving some problems in Saudi society.

However, the authors believe that the impact of social perception might not last for the coming generations.

References

1. Barnawi, E., & Ohkura, M. (2016). Evaluation of key emotional value that influence saudi women to purchase luxury fashion brands. In *Proceedings ICServ 2016*. 031, Tokyo.
2. Barnawi, E., & Ohkura, M. (2017). Key emotional value and Kawaii for Saudi women. In *Proceedings The 2nd Asian Conference on Ergonomics and Design (ACED2017)*. B5–2, Chiba.
3. Ministry of Economy Trade and Industry of Japan, Kansei Value Creation Initiative. Retrieved from http://www.meti.go.jp/english/information/downloadfiles/PressRelease/080620KANSEI.pdf.
4. Ministry of Economy Trade and Industry of Japan, Kansei Initiative from materialistic fulfillment to emotional fulfillment. Retrieved from http://www.meti.go.jp/english/policy/mono_info_service/mono/kansei2009/.
5. Clean Clothes Campaign, General Factsheet Garment Industry. (2015). Retrieved from http://www.cleanclothes.org/resources/publications/factsheets/general-factsheet-garment-industry-february-2015.pdf/view.
6. Rajasekera, J., & Karunasena, H. (2015). Apparel design optimization for global market: Kansei engineering preference model. *International Journal of Affective Engineering, 14*(2), 119–126.
7. Koma, K. (2013). Kawaii as represented in scientific research: The possibilities of kawaii cultural studies. *Hemispheres. Studies on Cultures and Societies , 28*, 103–116.
8. Alsubaie, H., Valenzuela, F. R., & Adapa, S. (2015). The advent of western-style shopping centres and changes in Saudi women's purchasing behaviour. In *Emerging Research on Islamic Marketing and Tourism in the Global Economy*.
9. Aqeel, A. (2012). Factors influencing Saudi women to purchase luxury fashion brand, Malaysia, *International Conference on Management, Behavioral Sciences and Economics Issues*.
10. Hofstede, G. (1994). *Management Scientists Are Human*. Netherlands: Management Science.
11. Hofstede, G. Retrieved from https://geert-hofstede.com/saudi-arabia.html.
12. Alamri, A., Cristea, A., & Al-Zaidi, M. (2014). Saudi Arabian cultural factors and personalised E-Learning, Spain, In *Proceedings of International Conference on Education and New learning Technologies*.
13. Hashizume, A., & Kurosu, M. (2013). Understanding user experience and artifact development through qualitative investigation: Ethnographic approach for human-cantered design, In *Proceedings of the 15th International Conference on Human-Computer Interaction 2013*.
14. Hayakawa, S., Ueda, Y., Go, K., & Takahashi, K. (2013). User research for experience vision. In *Proceedings of the 15th International Conference on Human-Computer Interaction 2013*.
15. Mayring, P. (2000). Qualitative Content Analysis, *Qualitative Social Research, 1*(2), Art.20.
16. Corbin, J., & Strauss, A. (2015). *Basics of qualitative research; techniques and procedures for developing grounded theory*. London: SAGE.

Chapter 11
Model of Kawaii Spoon Evaluations by Thai and Japanese

Tipporn Laohakangvalvit, Tiranee Achalakul and Michiko Ohkura

Abstract Affective values are critical factors for manufacturing in Japan. The significance of kawaii, which is a positive adjective that denotes such connotations as cute, lovable, and charming, continues to grow as an affective value, especially since it plays an important role in the success of such brands as Hello Kitty and Pokemon. Based on this success, we believe that kawaii will be a key factor for future product design. Kawaii culture is popular not only in Japan but also in many other Asian countries including Thailand. However, it remains unclear how closely the impressions of kawaii perceived by Thai people resemble those of Japanese people. Therefore, we compared 39 spoon designs based on Japanese and Thai feelings about kawaiiness and clarified the similarities and differences of these designs, which provided some design ideas for kawaii spoons. Since our results might not be applicable to evaluate general kawaii spoon designs, we constructed models of kawaii feelings with which to evaluate the kawaiiness of the spoon designs and clarified useful attributes.

Keywords Affective value · Kawaii · Spoon designs · Thai · Japanese

11.1 Introduction

This chapter is based on conference papers [1, 2] and their extensions. Kawaii culture is popular not only in Japan but it is also spreading around the world, especially in other Asian countries. Thailand is one country that has been strongly influenced by kawaii culture for several years. One research work [3] investigated how the Thai people, especially young females, have embraced it through such daily products as clothing and cosmetics. Moreover, the word "kawaii" itself is transliterated and frequently used in daily conversation by both Thai males and females. However, since various factors are impacting the shift of kawaii impressions away from its original

T. Laohakangvalvit (✉) · M. Ohkura
Shibaura Institute of Technology, Tokyo, Japan
e-mail: nb15505@shibaura-it.ac.jp

T. Achalakul
King Mongkut's University of Technology Thonburi, Bangkok, Thailand

© Springer Nature Singapore Pte Ltd. 2019
M. Ohkura (ed.), *Kawaii Engineering*, Springer Series on Cultural
Computing, https://doi.org/10.1007/978-981-13-7964-2_11

connotations, it remains unclear how much the impressions of kawaii perceived by Thai people resemble those of Japanese people. Identifying the similarities and the differences between the impressions in both countries will influence the design of future kawaii products that will serve the preferences of their respective consumers.

In the first part of this research, we experimentally observed kawaii feelings about spoon designs [1]. We employed 39 kawaii spoon designs and compared them based on the degree of kawaiiness felt by Japanese and Thai participants. Our experimental results clarified the similarities and the differences of kawaii preferences for spoon designs by gender and nationality. Based on the results, we suggested design ideas to spoon manufacturers in both countries. However, the applications of such suggestions are limited because they showed only examples of spoon designs without being extending to such further detailed analysis as suggestions about the attributes for kawaii spoon designs. Therefore, our results might not be applicable to evaluate general kawaii spoon designs.

In the second part of this research, we employed compared results of 39 spoon designs and their scores obtained from both Japanese and Thai participants [1]. We extracted a set of spoon attributes, constructed models with a machine learning algorithm, and clarified useful attributes for kawaii spoon designs [2].

11.2 Literature Survey

11.2.1 Affective Engineering and Kawaii Research

In Chap. 3, we studied various aspects of kawaii feelings. One research study explored the idea of designing kawaii products by examining kawaii attributes for shape, color, size, texture, and tactile sensation (Chaps. 2–5). Chapter 6 studied the preferences of kawaiiness among six illustrations by gender and age among Japanese participants and clarified that Japanese females have the strongest interest in kawaii.

As kawaii culture becomes more popular, the research focus on kawaii feelings is not limited to Japanese people. Chapter 10 observed the similarities and differences of kawaiiness by gender and age of Japanese and Saudi Arabians. Another research study [1] investigated kawaii culture in Thailand through various daily products preferred by Thai people, especially young females. Since the finding of that study resembles our previous study's conclusion that females tend to be more interested in kawaii, we extended our research to systematically study the similarities and differences by gender and between Japanese and Thai nationalities.

11.2.2   Feature Extraction Techniques

Feature extraction creates features or attributes that make machine learning algorithms work. In this study, we used spoon designs that were represented as images. Therefore, we needed to extract attributes from these images to identify the critical attributes for kawaii feelings by a machine learning algorithm.

Some previous research studies related to affective engineering used shape and color to address the extraction of useful information from images [4, 5]. Since we agree that shape and color are key components, we also chose to use them to extract attributes from our images of spoon designs.

There are various techniques for feature extraction. Traditional methods, which manually extract the attributes, often require expensive human labor because they require human observations. However, if the problem is not complicated, this method reduces learning time and cost. Another method exploits image-processing techniques to transform images into attributes represented in numerical forms. Since color is one of the most important aspects of images and has been successfully used in various applications [6], we employed such color spaces as RGB and HSV, to extract attributes from our images of spoon designs.

11.2.3   Machine Learning Techniques

Machine learning algorithms are widely used in many applications to learn about and make predictions about data. We used supervised learning to perform classification in which a computer is given inputs and the desired outputs, and the goal is to learn a general rule that maps the former to the latter.

Various approaches can be chosen for machine learning algorithms for classification, for example, an artificial neural network (ANN), a support vector machine (SVM), Bayesian networks, decision trees, and others. In this section, we discuss only SVM algorithms because they have been successfully used in many studies related to image-based classification [7, 8]. SVM classifiers are effective in high-dimensional feature spaces, especially when the training dataset is small [9]. They can efficiently perform linear or nonlinear classification using different kernels, for example, linear, polynomial, radial basis function (RBF), and sigmoid kernels. Usually, linear and RBF kernels are commonly used because they yield good results. One study showed that a linear kernel is a degenerated version of an RBF kernel [10]. Therefore, in many cases, RBF kernels are chosen.

11.3   Experimental Method

11.3.1   Collection of Spoon Designs

Our participants were female students at Tokyo Woman's Christian University who designed kawaii spoons using a spoon design layout provided by Aoyoshi Co., Ltd [11], a manufacturer of household goods. This layout consists of top and side views (Fig. 11.1). The women created a kawaii spoon by drawing on the given layout. We collected 182 spoon designs.

11.3.2   Preparation of Spoon Designs

We selected the best examples of spoon designs for our experiment. In this step, we only considered the top-view spoon designs because some side-view spoon designs were too difficult to draw and the actual design intention might have been misrepresented. Based on the practicality of using a spoon, we only considered handle designs.

We excluded 20 of 182 spoon designs based on the following criteria:

- Three designs with cartoon characters were removed to eliminate potential preference bias.
- 17 designs with non-surfaced decorations: This group included designs with items that were placed over the spoon's surface. They were excluded due to such practical issues as being difficult to use or to clean.

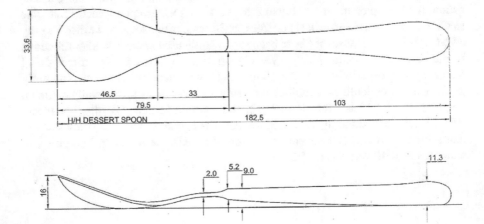

Fig. 11.1 Layout of spoon design (Aoyoshi Co., Ltd.)

We divided the remaining 162 spoon designs into groups based on the appearances or the shapes of the objects on the designs, such as flowers, hearts, smileys, stars, cats, and others. We counted the number of designs in each group and selected spoon designs from the three shape groups with the highest numbers of designs: the flower group (37 designs), the heart group (13 designs), and the smiley group (13 designs). We selected them as candidates for the experiment and excluded the other 99 designs.

11.3.3   Preliminary Experiment

We divided the remaining three designs into three shape groups. However, since the number of designs in the flower group was much larger than the other two groups, we performed a preliminary experiment to reduce the number of flower designs.

11.3.3.1   Participants

We recruited three female participants of different nationalities: Japanese, Thai, and Brazilian. All of them are interested in kawaii culture.

11.3.3.2   Preliminary Experimental Procedure

We created a questionnaire using Google Form. The participants answered it online using their own PCs. The questionnaire started with instruction that described how to evaluate the spoon designs. Then it advanced to the question parts, which included 37 questions, each of which showed a spoon design image and asked, "How kawaii is this spoon design?" (Fig. 11.2). The participants rated the designs on a 4-point scale: 0 (not at all), 1 (low), 2 (moderate), and 3 (high). After rating all 37 spoon designs, the participants clicked the "submit" button to save their questionnaire results in the system.

11.3.3.3   Preliminary Experimental Results

Based on the questionnaire results, we added the ratings for each spoon design from the ratings of all three participants and selected the 13 highest rated designs as more kawaii. We used these selected designs and the other designs in the heart and smiley groups in the experiment described in the next section.

Fig. 11.2 Screenshot of
questionnaire in preliminary
experiment

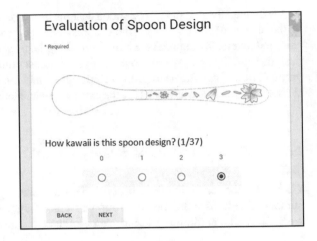

11.4 Experiment on Comparison of Spoon Designs

11.4.1 Spoon Designs Candidates

We obtained 39 spoon designs and divided them into three shape groups: 13 flower designs, 13 heart designs, and 13 smiley designs. However, since these designs were originally hand-drawn by students with different drawing ability, all 39 designs were graphically redrawn by a professional artist to remove any potential bias based on drawing quality (Fig. 11.3).

11.4.2 Comparison System

We modified a spoon comparison system from a previous scheme that evaluated kawaii illustrations (Chap. 6). This system collected the comparison results of kawaii spoon designs. As visual stimuli, the system used 39 graphical spoon designs that were displayed in pairs on a PC monitor (Fig. 11.4).

The following is the system's structure:

1. Top page: explanation of comparison method;
2. Consent form: brief explanation about the experiment and permission to use their data;
3. Input of participant's gender and nationality;
4. Instructions about answering the comparison result of the spoon designs;
5. Spoon design comparison: pairs of spoon designs were displayed for five seconds by a countdown timer. Then the participants are selected from three comparison results (more, less, or equally kawaii) using the keyboard's arrow keys; and

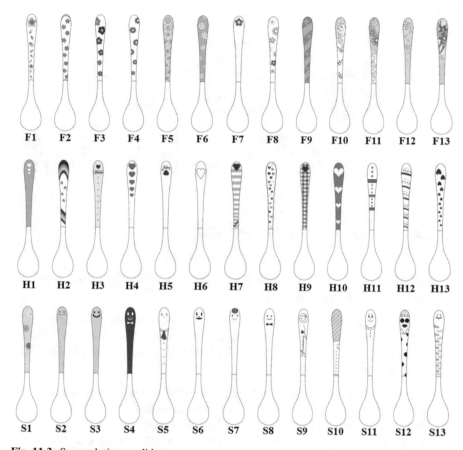

Fig. 11.3 Spoon design candidates

Fig. 11.4 Screenshot of spoon comparison system that displayed two spoon designs

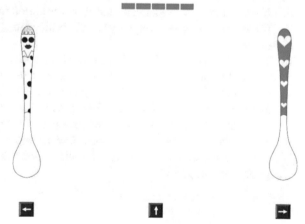

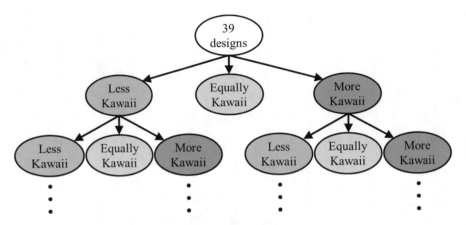

Fig. 11.5 Tree structure of quicksort-based comparison

6. Last page: the system stated that the comparison was finished and saved the participant's result in a database.

The comparison method used in the system was a quicksort algorithm (Fig. 11.5). Since a large number of spoon designs needed to be compared, this method reduced the comparison number and duration and still evaluated all of the spoon designs. The following steps explain our method:

1. A spoon design was randomly selected and shown on the left side as a pivot (P).
2. Another spoon design (Q) was selected and shown on the right side.
3. Participants compared a pair of spoon designs: P versus Q.
4. They answered by the keyboard's arrow key. Q was sorted into one of the following groups based on the answers.
5. If Q was deemed more kawaii than P, it was sorted into the more kawaii group.
6. If it was deemed less kawaii than P, it was sorted into the less kawaii group.
7. If P and Q were deemed equally kawaii, both were placed in the "equally kawaii" group.
8. The system repeated Steps 2–4 until all the spoon designs were compared.
9. Spoon designs were divided into three groups: more kawaii (A), less kawaii (B), and equally kawaii (C). Spoon designs in the less kawaii group were compared again through Steps 1–5, and followed by those in the more kawaii groups. For spoon designs in the equally kawaii group, the system stopped comparing them.
10. Step 6 was repeated until the more kawaii and less kawaii groups contained only one spoon design, denoting no other spoon designs remained in the same group for comparison.

11.4.3 Experimental Setup and Procedure

Our comparison system was displayed on a web browser with a 13-inch monitor and resolution of 3200 × 1800 pixels. The following are the experimental procedures:

1. Participants sat on a chair in front of a PC.
2. The experimenter showed the spoon comparison system.
3. Participants read the explanation and instruction on the display.
4. Participants submitted a consent form and agreed to cooperate in the experiment.
5. They compared the pairs of spoon designs shown on the PC display.
6. They input their answers about the comparison results.

11.5 Experimental Results

11.5.1 Participants

The experiment was performed with 40 volunteers, all of whom were university students in their 20s. They were equally divided into four groups by gender and nationality: ten Thai males, ten Thai females, ten Japanese males, and ten Japanese females.

11.5.2 Comparison Results

From the comparison results of each participant, we sorted all 39 spoon designs and ranked them using the method described below.

Using a tree structure for the sorting result, the spoon designs in the bottom more kawaii group were ranked as the most kawaii or the top rank. The spoon designs in the bottom less kawaii group were ranked as the least kawaii. Two or more spoon designs in the equally kawaii group had the same rank.

From the rankings, we calculated the scores of all the spoon designs. The top-ranked score was 39. If the rank was worse, the score was lowered. An example of the method that ranked and scored the spoon designs is shown in Fig. 11.6. We used the scores to compare the spoon designs and to analyze the results in the following sections.

11.5.2.1 Comparison of Three Shape Groups of Spoon Designs

We compared the scores of the spoon designs that were divided into flower, heart, and smiley groups. From the one-factor ANOVA result, shape had a significant main

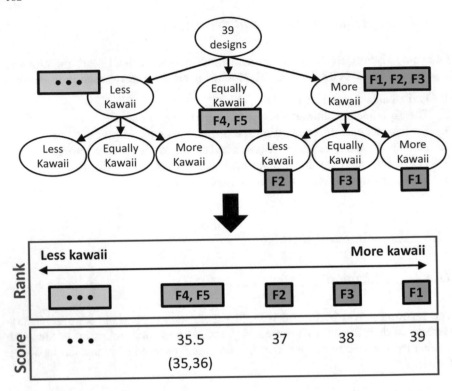

Fig. 11.6 Example of ranking and scoring spoon designs

Fig. 11.7 Average scores of spoon designs divided into flower, heart, and smiley groups for all participants

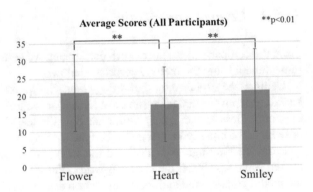

effect (p < 0.01). Tukey post hoc tests revealed statistically significant differences in the average scores between the flower and heart groups (p < 0.01) and between the smiley and heart groups (p < 0.01) (Fig. 11.7). This result indicates that the average scores of the spoon designs in the flower and smiley groups were higher than the heart group.

Table 11.1 Comparison results between participant groups by Spearman's rank-order correlation

	Thai males	Thai females	Japanese males	Japanese females
Thai males	–	0.329*	0.649**	0.294
Thai females		–	0.104	0.342*
Japanese males			–	0.051
Japanese females				–

** Correlation is significant at 0.01 level
* Correlation is significant at 0.05 level

11.5.2.2 Correlation Analysis

We performed a correlation analysis on the scores of the 39 spoon designs among the four participant groups using Spearman's rank-order correlation. Spearman's correlation coefficient (r_s) measures the strength and the direction of the monotonic relationships between two ranked variables. r_s ranges from -1 to $+1$, where -1 indicates a perfect negative association of the ranks, $+1$ indicates a perfect positive association of the ranks, and 0 indicates no association.

We compared the scores between four pairs of participant groups to clarify the similarities and differences between gender and nationality. The r_s values of the four comparisons are shown in Table 11.1.

To compare the genders, we compared the r_s of the Thai versus Japanese males ($r_s = 0.649$, $p < 0.01$) and the Thai versus Japanese females ($r_s = 0.342$, $p < 0.05$). The r_s of the males was higher, which indicated a stronger correlation between males than females.

For a comparison between nationalities, we compared the r_s of the Thai males versus females ($r_s = 0.329$, $p < 0.01$) and the Japanese males versus females ($r_s = 0.051$). The r_s of the Thais was higher, which indicates a stronger correlation between Thais than Japanese.

From the correlation analysis results, Thai males versus Japanese males had a strong correlation, which indicates that they had a similar ranking tendency.

These combinations had moderate correlations: Thai females versus Japanese females, Thai males versus Thai females, and Thai males versus Japanese females.

The combinations of Thai females versus Japanese males and Japanese males versus Japanese females had very weak or no correlations.

11.5.2.3 Comparison of Individual Spoon Designs

We compared the scores of each spoon design to clarify the similarities and the differences among the four participant groups. Figure 11.8 shows the average scores of their 39 spoon designs.

First, we analyzed their differences. From paired t-test results, we identified the following significant differences in the scores for some spoon designs:

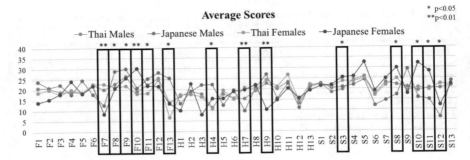

Fig. 11.8 Average scores of 39 spoon designs among four participant groups. Spoon designs with square brackets (F7, F8, F9, F10, F11, F13, H4, H7, H9, S3, S8, S10, S11, S12) indicated significant differences among scores

- Thai males versus Japanese males: designs H4 and S12;
- Thai females versus Japanese females: designs F7, F10, and H9;
- Thai males versus Thai females: designs F9, F11, F13, H7, and S3;
- Japanese males versus Japanese females: designs F8, H9, S8, S10, and S11.

Based on this result, we eliminated these 14 spoon designs. 25 spoon designs remained without any statistically significant differences between any pairs of the four participant groups. Each of these spoon designs had a similar ranking tendency for all the participants. Therefore, we analyzed their similarities by gender and nationality. A one-factor ANOVA showed a significant main effect of the spoon designs ($p < 0.01$). Tukey post hoc tests revealed significant differences between the scores of the following spoon designs:

- For all participants:

  - S5 > H1, H3, H5, H6, H10, H12, S6
  - F12 > H1, H3, H5, H12, S6
  - S4 > H1, H12
  - H11, S1, S9, S13 > H12;

- For Japanese females:

  - S5 > F1, F2, H1, H3, H5, H10, H12
  - S4, S7 > H3;

- For Japanese males:

  - S9 > H5, H12, S6
  - F12 > H12;

- For Japanese:

  - S5 > H1, H3, H5, H10, H12, S6
  - F12, S4 > H1;

Gender/Nationality	Spoon design	
	Top 3	Bottom 3
All participants	S4, S5, F12	H1, H5, H12
Japanese females	S4, S5, S7	H1, H3, H10
Japanese males	S5, S9, F12	H5, S6, H12
All Japanese	S4, S5, F12	H1, H5, H12
All Thai	S4, S5, F12	H1, H6, H12
Thai females	–	–
Thai males	–	–
Females	S4, S5, H11	H1, H3, H12
Males	S5, S9, F12	H1, H5, H12

Table 11.2 Top and bottom three spoon designs for each participant group

- For Thai:

  - S5 > H12;

- For females:

  - S5 > F1, H1, H3, H12
  - S4 > H3;

- For males:

  - F12 > H1, H5, H12, S6
  - F1, S5, S9 > H12.

- For both Thai males and females, we found no significant differences in any pairs of spoon designs.

From a comparison of the scores, we listed the top and bottom three spoon designs for each participant group (Table 11.2).

11.6   Method for Model Construction

11.6.1   Overview of Model Construction

Model construction was divided into two phases: training and validation (Fig. 11.9). The training phase consisted of these following two steps:

1. Preparation of raw data (i.e., images of spoon designs) for feature extraction.
2. Training the models using feature vectors (i.e., attributes of spoon designs) and labels as input to the SVM classification algorithm.

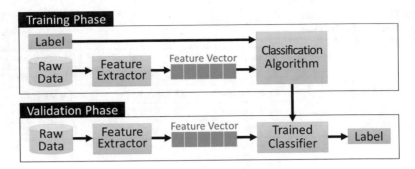

Fig. 11.9 Overall procedure of model construction

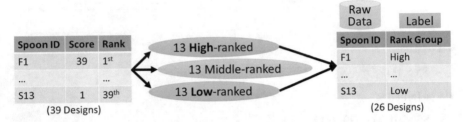

Fig. 11.10 Procedure of dataset preparation

The validation phase confirmed the classification accuracy of the trained classifiers from the training phase. The details of each phase are described in the following sections.

11.6.2 Dataset Preparation

To construct our model, we employed spoon designs as target products that had already been evaluated. Their attributes, which were not so complexed, were appropriate for the first trial of model construction. We used the average scores of 39 spoon designs from the following groups of participants: all, male, and female. Based on the average scores, the ranks of the 39 spoon designs were calculated and divided into three groups: high, middle, and low ranked (Fig. 11.10). We only used the spoon designs and their order in the high-ranked group (top 13 ranks) and the low-ranked group (bottom 13 ranks) as the dataset. This procedure of dataset preparation was repeated to prepare three datasets for all, male, and female participants.

Table 11.3 List of attributes defined by manual observation method

Attribute	Value
Shape-based attributes	
Shape	Flower/Heart/Smiley
Is the shape repeated?	Yes/No
Number of shapes	(Integer)
Are there other shapes?	Yes/No
Background pattern	None/Plain color/ Patterned
Color-based attributes	
Does the shape have color?	Yes/No
Does the shape have a color gradient?	Yes/No
Does the background have color?	Yes/No
Does the background have a color gradient?	Yes/No
Total number of colors	1/2/3/More than 3

11.6.3 Feature Extraction

We defined 33 physical attributes and calculated their values for the 39 spoon designs. To define the attributes, we extracted the features from the images of the spoon designs into categorical or numerical forms that were required as input to train the model using the SVM algorithm in the next step. We performed feature extraction using the following two methods.

11.6.3.1 Manual Observation

This method manually observed the appearance of the spoon designs. Based on shape and color, we defined the ten attributes shown in Table 11.3.

11.6.3.2 Image Processing

Defining the attributes using only the manual observation might not be sufficiently efficient. Thus, we also employed image processing techniques to calculate the pixel values and the number of pixels on the images of the spoon designs. To calculate those values, we divided each image of spoon designs into the following three areas:

- Total area (Fig. 11.11a): entire image;
- Shape area (Fig. 11.11b): area that represents flower, heart, or smiley shapes;
- Background area (Fig. 11.11c): area other than the shape area.

Based on shape and color, we defined the 23 attributes shown in Table 11.4.

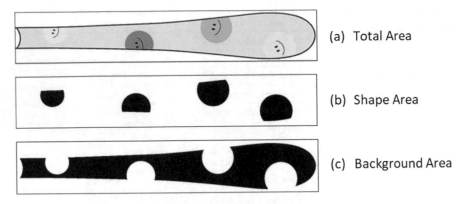

Fig. 11.11 **a** Original spoon design with smiley shape; **b** black pixels represent area of smiley shape; **c** black pixels represent background area

For the shape-based attributes, we calculated the ratio of pixels among the areas. For example, to obtain the "ratio of pixels in the shape area to the total area" of the spoon design in Fig. 11.11, we calculated the ratio of black pixels that represent smiley shape (B) to all pixels (A).

For the color-based attributes, we employed two color spaces, Red-Green-Blue (RGB) and Hue-Saturation-Value/Brightness (HSV), and calculated the average pixel values in the shape, background, and total areas.

From both feature extraction methods, we defined a total of 33 attributes of spoon designs and used the extracted attributes in the next step.

11.6.4 Model Training

We combined the spoon designs and their ranking groups (Sect. 11.6.1) and the defined attributes (Sect. 11.6.2) to create a dataset (Fig. 11.12). Each spoon design consisted of its ranking group (high or low) and a set of 33 attributes with corresponding values as a feature vector. We prepared datasets for the all, male, and female participants. The 26 spoon designs and their ranked groups were different among the three participant groups.

We used the SVM (RBF) algorithm to train the classifiers. We performed threefold cross-validation in which 70% of the dataset was used as a training set. Therefore, the classifier was trained with different data for three iterations (Fig. 11.13).

Table 11.4 List of attributes defined by image processing method

Attribute	Category
Shape-based attributes	
Ratio of shape area to total area	(Integer, Range = 0–100)
Ratio of background area to total area	(Integer, Range = 0–100)
Ratio of colored area to total area (Note that white was ignored)	(Integer, Range = 0–100)
Ratio of shape area to colored area	(Integer, Range = 0–1)
Ratio of background area to colored area	(Integer, Range = 0–1)
Color-based attributes	
Average R in total area	(Integer, Range = 0–255)
Average G in total area	(Integer, Range = 0–255)
Average B in total area	(Integer, Range = 0–255)
Average H in total area	(Integer, Range = 0–360)
Average S in total area	(Integer, Range = 0–100)
Average V in total area	(Integer, Range = 0–100)
Average R in shape area	(Integer, Range = 0–255)
Average G in shape area	(Integer, Range = 0–255)
Average B in shape area	(Integer, Range = 0–255)
Average H in shape area	(Integer, Range = 0–360)
Average S in shape area	(Integer, Range = 0–100)
Average V in shape area	(Integer, Range = 0–100)
Average R in background area	(Integer, Range = 0–255)
Average G in background area	(Integer, Range = 0–255)
Average B in background area	(Integer, Range = 0–255)
Average H in background area	(Integer, Range = 0–360)
Average S in background area	(Integer, Range = 0–100)
Average V in background area	(Integer, Range = 0–100)

	Raw Data	Label	Set of Attributes (Feature Vector)			
	Spoon ID	Rank Group	Attribute 1	Attribute 2	...	Attribute 33
13 Spoon Designs (High-ranked groups)	F1	High	...	...	...	...
	...	High	...	...	...	...
13 Spoon Designs (Low-ranked groups)	S13	Low	...	...	...	...
	...	Low	...	...	...	...

Fig. 11.12 Procedure of training a classifier using prepared dataset

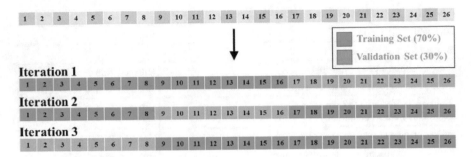

Fig. 11.13 Training and validation sets for each iteration of threefold cross-validation

Table 11.5 Classification accuracy of threefold cross-validation divided by participant groups

Participant group	Classification accuracy (%)			
	Iteration#1	Iteration#2	Iteration#3	Average
All participants	62.5	62.5	70.0	65.0
Males	75.0	62.5	80.0	72.5
Females	50.0	37.5	50.0	45.8

11.6.5 Model Validation

As previously mentioned, we performed a threefold cross-validation. For each iteration, the trained classifier was validated using the remaining 30% of the dataset. From the classification results, we calculated the accuracy of each trained classifier from the number of correctly classified data. If all of the input and predicted ranks were matched, the accuracy was 100%.

11.7 Results of Model Construction

11.7.1 Classification Results

From the prediction phase, we obtained the classification accuracy of the trained classifiers for the dataset of the all, male, and female participants (Table 11.5). Then, we calculated the average accuracy of the three iterations for each participant group. The average accuracies for the all, male, and female participants were 65.0%, 72.5%, and 45.8%, respectively.

The average accuracy for the female participants was lower than that of the male participants, reflecting our previous research (Chap. 6) where female participants had a greater variety of kawaii selections than males.

Table 11.6 Important attributes of the three iterations for males

Iteration#1		Iteration#2		Iteration#3	
Attribute	RI	Attribute	RI	Attribute	RI
Total number of colors (1)	0.049	Total number of colors (1)	0.037	Total number of colors (1)	0.067
Shape (2)	0.040	Are there other shapes?	0.033	Is the shape repeated?	0.048
Is the shape repeated?	0.036	Background pattern	0.032	Shape (2)	0.047
Ratio of shape area to total area	0.034	Average S in background area	0.032	Background pattern	0.042
Ratio of background area to total area	0.034	Average S in shape area	0.031	Does the background have color?	0.038
Average R in background area	0.032	Average H in shape area	0.031		
Does the shape have a color gradient?	0.032	Shape (2)	0.031		
Average V in background area	0.032	Average R in total area	0.031		
Average S in background area	0.032				
Average R in shape area	0.031				

11.7.2   Results of Effective Attributes

Since male and female participants had differences in their kawaii selection, we analyzed the results of the effective attributes by dividing the results by gender.

The model's effective attributes were indicated by the Relative Importance (RI), which is the values obtained from the SVM results. We set the RI threshold to 0.03. For each iteration, we obtained effective attributes and their RIs and selected the common attributes from the three iterations in which the attributes with RIs equal to or over 0.03 were included.

Male participants The effective attributes of the three iterations for the male participants are shown in Table 11.6. From the selection of the common attributes among three iterations, we found two effective attributes: (1) total number of colors and (2) shape.

As shown in Table 11.7, we further analyzed the values of the two effective attributes for the high- and low-ranked groups based on the ranks of the spoon designs in both groups. For the high-ranked group, the spoon designs with many colors (more than three) in the flower or smiley shapes tended to have high ranks, indicating that they are more kawaii. For low-ranked group, the spoon designs with fewer colors

Table 11.7 Values of important attributes for high- and low-ranked groups for male participants

Attribute	Rank Group	
	High ranked	Low ranked
Total number of colors	More colors	Fewer colors
Shape	1. Flower 2. Smiley	Heart

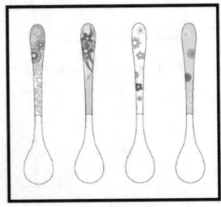 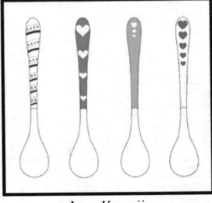

More Kawaii   Less Kawaii

Fig. 11.14 Examples of more and fewer kawaii spoon designs for males

(fewer than three) in the heart shape tended to have low ranks, indicating that they are less kawaii. Examples of more and fewer kawaii spoon designs are shown in Fig. 11.14.

Female participants We analyzed the result of the effective attributes using the same method as that of the male participants described above. The effective attributes of the three iterations for females are shown in Table 11.8. From the selection of common attributes, shape was the only effective attribute.

We further analyzed the values of the effective attributes for the high- and low-ranked groups (Table 11.9). For the high-ranked group, spoon designs with flower or smiley shapes tended to have higher ranks, indicating that they are more kawaii. For the low-ranked group, spoon designs with a heart shape tended to have lower ranks, indicating that they are less kawaii.

Table 11.8 Important attributes of the three iterations for females

Iteration#1		Iteration#2		Iteration#3	
Attribute	RI	Attribute	RI	Attribute	RI
Shape	0.047	Is the shape repeated?	0.059	Shape	0.042
Average R in shape area	0.035	Shape	0.057	Background pattern	0.032
Average V in shape area	0.034	Average S in background area	0.036	Does the background have color?	0.032
Ratio of shape area to total area	0.033	Average B in total area	0.032	Ratio of shape area to total area	0.032
Ratio of background area to total area	0.033	Average G in total area	0.032	Ratio of background area to total area	0.032
Background pattern	0.033	Average H in total area	0.031	Is the shape repeated?	0.031
Does the background have color?	0.033	Average H in background area	0.031	Average S in shape area	0.031
Average S in shape area	0.031	Average S in total area	0.031		
Average H in Background	0.031				
Total number of colors	0.031				

Table 11.9 Values of important attributes for high- and low-ranked groups for female participants

Attribute	Rank group	
	High ranked	Low ranked
Shape	1. Smiley 2. Flower	Heart

11.8 Conclusion

We developed a spoon comparison system and experimentally collected comparison results of 39 spoon designs based on kawaiiness between genders and nationalities (Sects. 11.3–11.5).

For kawaii spoons, all participants preferred flower and smiley designs over heart designs. Males tended to prefer flower designs and females tended to prefer smiley designs. The kawaii selection tendency between Japanese and Thai participants was similar. Considering each spoon design, we found that more than half had no differences in kawaii selection. Based on the results, we can provide suggestions to spoon manufacturers about kawaii designs for Japanese and Thai consumers.

Then, we performed further analysis to model kawaii feelings to give more general suggestions for kawaii spoon designs (Sects. 11.6–11.7). We ranked 39 spoon designs based on the evaluation results and defined 33 attributes based on shape and color to construct models of kawaii feelings for them. The models were constructed using the SVM algorithm, which classified the kawaiiness of the spoon designs. From the classification results, the model's accuracy for females was lower than that of males, indicating that females had a greater variety of kawaii selection than males. Perhaps manufacturers should heed genders when they are designing kawaii spoons. Finally, we clarified the effective attributes for kawaii spoon designs for males (total number of colors and shape) and females (shape). For spoon manufacturers, these attributes should be taken into account when designing kawaii spoons.

Acknowledgements Part of this work was supported by Grant-in-Aid Scientific Research Number 26280104. We thank the students of the Tokyo Woman's Christian University, King Mongkut's University of Technology Thonburi, and the Shibaura Institute of Technology for their participation.

References

1. Laohakangvalvit, T., & Ohkura, M. (2017). Comparison of spoon designs based on Kawaiiness between genders and nationalities. In *Proceedings of the International Symposium of Affective Science and Engineering (ISASE 2017)* (pp. 1–7), Tokyo, Japan.
2. Laohakangvalvit, T., Achalakul, T., & Ohkura, M. (2017). A Proposal of model of Kawaii feelings for spoon designs. In *Proceedings of the International Conference on Human-Computer Interaction (HCI 2017)* (pp. 687–699). Vancouver.
3. Toyoshima, N. (2015). Kawaii Fashion in Thailand. The Consumption of Cuteness from Japan. *Journal of Asia-Pacific Studies (Waseda University), 24*, 191–210.
4. Chen, Y., Sobue, S., & Huang, X. (2009). KANSEI based clothing fabric image retrieval. In *Proceedings of the Second International Workshop on Computational Color Imaging* (pp. 71–80). Suzhou.
5. Isomoto, Y., Yoshine, K., Yamasaki, H., & Ishii, N. (1999). Color, shape and impression keyword as attributes of paintings for information retrieval. In *Proceedings of the IEEE International Conference on Systems, Man, and Cybernetics* (pp. 257–262). Tokyo.
6. Acharya, T., & Ray, A. K. (2005) *Image processing: Principles and applications*. Wiley-Interscience Publication.
7. Bejarano, A. M., Calvo, A. F., Henao, C. A. (2016). Supervised learning models for control quality by using color descriptors: A study case. In *Proceedings of the XXI Symposium on Signal Processing, Images and Artificial Vision* (pp. 1–7). Bucaramanga.
8. Wong, W., & Hsu, S. (2006). Application of SVM and ANN for image retrieval. *European Journal of Operational Research, 173*(3), 938–950.
9. Vapnik, V. (2000). *The nature of statistical learning theory. Information science and statistics*. Springer.
10. Keerthi, S. S., & Lin, C. (2003). Asymtotic behaviors of support vector machine with gaussian kernel. *Neural Computation, 15*(7), 1667–1689.
11. Aoyoshi Co., Ltd. http://www.aoyoshi.co.jp.

Chapter 12
Model of Kawaii Cosmetic Bottle Evaluations by Thai and Japanese

Tipporn Laohakangvalvit, Tiranee Achalakul and Michiko Ohkura

Abstract Affective values are critical factors for manufacturing in Japan. Kawaii, an affirmative adjective that denotes such positive meanings as cute or lovable, has become even more critical as an affective value and plays a leading role in the world-wide success of many products, such as Hello Kitty and Pokemon. Based on this success, we believe that kawaii will be a key factor in future product design. In our previous research, we proposed models of kawaii feelings for spoon designs and extracted the attributes of such designs and constructed models using the Support Vector Machine (SVM) algorithm. In this research, we used the Deep Convolutional Neural Network (CNN) algorithm because it can perform classification using images as input and studied the kawaiiness of cosmetic bottles. Then, we evaluated the candidates of effective attributes with our model to increase the kawaiiness of cosmetic bottles. Finally, we clarified the relationship among kawaii feelings, attributes, and eye movement indexes obtained from our previous research, and the prediction results of our constructed model. Our results clarified the effective attributes for increasing kawaiiness and the effectiveness of our constructed model to evaluate the kawaiiness of cosmetic bottles.

Keywords Affective value · Cosmetic bottles · Deep convolutional neural network (CNN) · Eye tracking · Kawaii · Product attribute

12.1 Introduction

This chapter is based on conference papers [1, 2], a manuscript [3], and their extensions. Research studies have explored kawaii attributes for designing kawaii products such as shape, color, size, texture, and tactile sensation (Chaps. 2–5). In Chap. 11, we collected the comparison results of spoon designs based on kawaiiness and extracted

T. Laohakangvalvit (✉) · M. Ohkura
Shibaura Institute of Technology, Tokyo, Japan
e-mail: nb15505@shibaura-it.ac.jp

T. Achalakul
King Mongkut's University of Technology Thonburi, Bangkok, Thailand

their attributes based on shape and color. The comparison results and attributes were used to construct models of kawaii feelings for spoon designs. We used the Support Vector Machine (SVM) algorithm to classify the kawaiiness and constructed models for spoon designs. Finally, we clarified effective attributes to design kawaii spoons.

Moreover, research studies have systematically studied the kawaii feelings evoked by kawaii attributes in which biological signals were employed including heartbeat and brain waves (Chap. 5). However, even though eye tracking has not yet scrutinized kawaii feelings, it has been widely used in various research fields, including cognitive and experimental psychology, human–computer interaction, and product development. Eye tracking can recognize human emotional states and preferences. Our previous research (Chap. 6) employed it to evaluate kawaii feelings to investigate the relationship between kawaii feelings and eye movements. As a result, we clarified the relationships and identified eye movement indexes related to kawaii feelings.

In this research, we targeted cosmetic bottles. Research has studied the factors that influence the buying behaviors of cosmetic products and concluded that attractive packaging of perfumes [4] and cosmetic products [5] is one highly prioritized factor that influences purchase.

This research is divided into four main parts. The first part (Sect. 12.2) includes a collection of the evaluation data of cosmetic bottles. The second part (Sect. 12.3) includes a model of kawaii feelings for cosmetic bottles using the data collected from Sect. 12.2. To construct our model, the SVM algorithm that we previously used in Chap. 11 has a limitation: feature extraction is required to prepare the dataset as input. Thus applying this method might be considerably difficult for products that have unknown sets of features. In contrast, a Deep Convolutional Neural Network (CNN) algorithm has been successfully applied to various domains especially detection, segmentation, and recognition of images [6, 7]. One advantage is that feature extraction is not required, as in SVM, because it can perform classification using images as input. Based on its success in various studies, we used this method in this research.

The third part (Sect. 12.4) evaluates the attributes using our constructed model. Since our Deep CNN model cannot generate the candidates of effective attributes (as done by the SVM model), we developed a new method to obtain these attributes. Finally, the last part (Sect. 12.5) includes a clarification of the effective attributes for kawaii cosmetic bottles using our model and eye movement indexes obtained from our previous research in Chap. 6.

12.2 Data Collection of Evaluation of Kawaiiness for Cosmetic Bottles

In the next step, we collected cosmetic bottle images and evaluated their kawaiiness to gather data for constructing our model. The details are described in the following sections.

Fig. 12.1 Example of cosmetic bottle images (photos by www.pngimg.com, licensed under Creative Commons CC BY-NC 4.0)

12.2.1   Data Collection

We collected 1,048 online images of cosmetic bottles, all of which are actual products that are currently on sale (Fig. 12.1).

12.2.2   Experimental Setup and Procedure

We experimentally evaluated the kawaiiness of cosmetic bottle images. Since the number of images was large, we built a questionnaire system to facilitate our evaluations. The system showed each image on a screen one at a time. We used a 13.3-inch laptop with a resolution of 3200 × 1800 pixels.

The following is the experimental procedure:

1. Top page: explanation of experiment;
2. Consent form: brief explanation about experiment and permission to use their data;
3. Recording of gender and age;
4. Explanation of evaluations: each cosmetic bottle image is displayed, and participants evaluated it as kawaii or not-kawaii using the keyboard's left or right arrow keys;
5. Evaluation of cosmetic bottle images: 1,048 images are displayed and evaluated by each participant.

6. After the participants submitted their questionnaires, the evaluation results were saved in a database.

12.2.3 Experimental Results

We performed the experiment with the following participants.

- 15 Thai volunteers: ten females (average age $= 28.1$, SD $= 2.2$) and five males (average age $= 28.6$, SD $= 1.9$).
- 20 Japanese volunteers: ten females (average age $= 21.8$, SD $= 0.7$) and ten males (average age $= 21.7$, SD $= 0.7$).

We categorized the number of cosmetic bottle images evaluated by each participant as kawaii or not-kawaii. Based on the evaluations, all of the images were divided into kawaii or not-kawaii groups. The results of the Thai participants are shown in Table 12.1, and those of the Japanese participants are shown in Table 12.2.

To provide balanced data between the kawaii and not-kawaii groups for the model construction in the next step, we calculated the ratio of the number of images between the kawaii (A) and not-kawaii (B) groups. Then we defined whether the ratio of each participant was balanced. If the ratio was equal or close to 1:1, the data were

Table 12.1 Evaluation results of cosmetic bottle images of Thai participants

Participant ID (gender)	Number of images		Ratio (A:B)	Balanced data
	Kawaii (A)	Not-kawaii (B)		
P01 (F)	449	599	1:1.3	Yes
P02 (F)	627	421	1.5:1	Yes
P03 (F)	322	726	1:2.3	Yes
P04 (F)	136	912	1:6.7	No
P05 (F)	386	662	1:1.7	Yes
P06 (F)	72	976	1:13.6	No
P07 (F)	116	932	1:8	No
P08 (F)	73	975	1:13.4	No
P09 (F)	123	925	1:7.5	No
P10 (F)	698	350	2:1	Yes
P11 (M)	79	969	1:12.3	No
P12 (M)	623	425	1.5:1	Yes
P13 (M)	191	857	1:4.5	No
P14 (M)	273	775	1:2.8	Yes
P15 (M)	672	376	1:1.8	Yes

Table 12.2 Evaluation results of cosmetic bottle images of Japanese participants

Participant ID (gender)	Number of images		Ratio (A:B)	Balanced data
	Kawaii (A)	Not-kawaii (B)		
P01 (F)	241	807	1:3.3	Yes
P02 (F)	180	868	1:4.8	No
P03 (F)	611	437	1.4:1	Yes
P04 (F)	446	602	1:1.3	Yes
P05 (F)	149	899	1:6	No
P06 (F)	424	624	1:1.5	Yes
P07 (F)	676	372	1.8:1	Yes
P08 (F)	175	873	1:5	No
P09 (F)	248	800	1:3.2	Yes
P10 (F)	519	529	1:1	Yes
P11 (M)	731	317	2.3:1	Yes
P12 (M)	143	905	1:6.3	No
P13 (M)	468	580	1:1.2	Yes
P14 (M)	435	613	1:1.4	Yes
P15 (M)	532	516	1:1	Yes
P16 (M)	459	589	1:1.3	Yes
P17 (M)	290	758	1:2.6	Yes
P18 (M)	218	830	1:3.8	No
P19 (M)	124	924	1:7.5	No
P20 (M)	500	548	1:1.1	Yes

considered balanced. We selected eight Thai and 14 Japanese participants for further analysis.

Next, we considered the consistency of the evaluation results among participants for both nationalities using a two-step cluster analysis. When the evaluation results among participants are similar, we clustered them into the same group.

The Thai participants were clustered into two groups of four participants. However, using the results of four participants for model construction was too few. Therefore, we used the evaluation results of all eight participants for the model construction in the next step.

The Japanese participants were clustered into two groups: 11 participants in the first group and the remaining three participants in the second group. We used only the majority of the evaluation results (11 participants) for model construction in the next step.

12.3 Construction of Model of Kawaii Feelings for Cosmetic Bottles

We proposed a model of kawaii feelings for cosmetic bottles as a classifier of cosmetic bottle images by kawaiiness (kawaii or not-kawaii). In this section, we describe the procedure to prepare the dataset, construct our model, and validate it.

12.3.1 Dataset Preparation

For the Thai participants, we used as the dataset the evaluated cosmetic bottle images of eight participants from the experiment described in the previous section. It consisted of 8,384 images that were classified into kawaii (4,050 images) and not-kawaii (4,334 images) groups.

Since the results were from eight participants, there were eight identical images in the dataset. However, they could be classified into either kawaii or not-kawaii groups based on the evaluation of each participant, which might be inconsistent. However, inconsistency is also important to indicate how much each image contributed to the kawaii or not-kawaii groups. Therefore, we used all the images for model construction.

To prevent model overfitting problems [8], we used k-fold cross-validation (k = 8). Since the dataset contained the results of eight participants, we selected eightfold so that the dataset could be divided into eight subsets without repetition of identical images in the same subset. Our method for dividing the dataset is described in the following steps (Fig. 12.2):

1. For each participant, the images were equally divided into eight subsets for both the kawaii and not-kawaii groups.
2. To avoid repetition of the same images in the same subsets, some images in a subset were swapped with others of the same participant until all the images in each part were different.
3. For all participants, seven subsets (v1 to v7) were used as training data. The remainder (v8) was used as the testing data.
4. The training data from the kawaii and not-kawaii groups were combined into a *training set*, which contained 87.5% of all the images.
5. The testing data from the kawaii and not-kawaii groups were combined into a *testing set*, which contained the remaining 12.5% of the images.

Note that each subset was rotated and used once as the testing data in Step 2. Thus we prepared a total of eight different training and testing sets.

For the Japanese participants, we used the evaluation results of 11 participants from the previous step as the dataset, which consisted of 11,528 images that were classified into kawaii (5,354 images) and not-kawaii (6,174 images) groups.

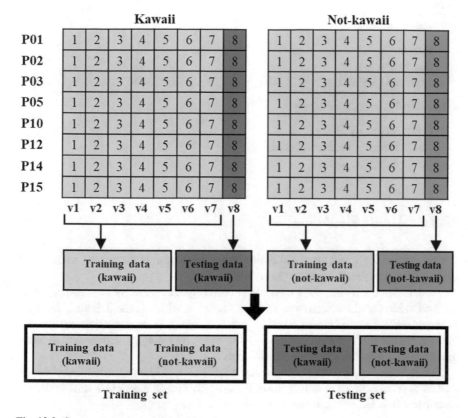

Fig. 12.2 Steps to prepare training and testing sets

By using the same method as that of the Thai participants, we prepared the dataset for the k-fold cross validation (k = 11). It was divided into 11 subsets (11 different training and testing sets). We changed the number of folds (k) equal the number of subsets for cross-validation, to ensure that all 1,048 images equally appear once in every subset.

12.3.2   Model Training

From the data preparation step, we obtained two datasets: Thai participants and Japanese participants. For the Thai dataset, we used eight training sets to train eight different models for cross-validation. Similarly, for the Japanese dataset, we used 11 training sets to train 11 different models.

To train the model, we employed a Deep Convolutional Neural Network algorithm called GoogLeNet with Batch Normalization. The following are the parameter settings.

- Training settings:

  - Epoch: 50
  - Batch size: 64
  - Learning rate: 0.01

- Image preprocessing settings:

  - Color mode: RGB
  - Image resizing method: Squash
  - Image flipping: Yes

12.3.3 Model Validation

We performed eightfold cross-validation using the above eight trained models and eight testing sets that correspond to each model.

To determine whether each image was correctly classified by the model, we set a cutoff value (or classification probability) at 50%, which is the default value for a binary classifier. For example, if an image's probability in the kawaii group exceeds 50%, it is classified as kawaii. On the other hand, any images with a classification probability of less than 50% are classified as not-kawaii.

For each model, we used a confusion matrix (Fig. 12.3) to measure the performance of a classification model using three performance metrics. The formulas to calculate and interpret each metric are shown as follows:

- Accuracy = (TP + TN)/(TP + TN + FP + FN); Ability to differentiate between kawaii and not-kawaii;
- Sensitivity = TP/(TP + FN); Ability to correctly determine kawaii case;
- Specificity = TN/(TN + FP); Ability to correctly determine not-kawaii case.

Fig. 12.3 Confusion matrix

		Actual condition	
		Kawaii	**Not-kawaii**
Predicted condition	**Kawaii**	True positive (TP)	False positive (FP)
	Not-kawaii	False negative (FN)	True negative (TN)

12.3.4 Results of Model Validation

For the results of the Thai dataset (Table 12.3), all three performance metrics were in certain ranges, which indicate its consistency. This result ensured that the robustness to construct a final model using this dataset.

For the Japanese dataset results (Table 12.4), even though the ranges of sensitivity and specificity had variability, the range of accuracy had consistency that resembled that of the Thai participants.

Table 12.3 Classification results of all models of Thai participants

Model No.	Number of correctly classified images/Number of all images to be classified		Accuracy (%)	Sensitivity (%)	Specificity (%)
	Kawaii	Not-kawaii			
1	255/507	398/542	62	50	73
2	265/507	417/541	65	52	77
3	262/506	402/542	63	52	74
4	276/506	399/542	64	55	74
5	268/506	422/542	66	52	78
6	246/506	423/542	64	49	78
7	280/506	384/541	63	55	71
8	250/506	407/541	63	49	75

Table 12.4 Classification results of all models of Japanese participants

Model No.	Number of correctly classified images/Number of all images to be classified		Accuracy (%)	Sensitivity (%)	Specificity (%)
	Kawaii	Not-kawaii			
1	277/487	399/562	64	57	71
2	268/487	393/562	63	55	70
3	303/487	357/562	63	62	64
4	253/487	435/561	66	52	78
5	207/487	454/561	63	43	81
6	235/487	424/561	63	48	76
7	170/487	472/561	61	35	84
8	255/487	407/561	63	52	73
9	288/486	369/561	63	59	66
10	212/486	463/561	64	44	83
11	200/486	461/561	63	41	82

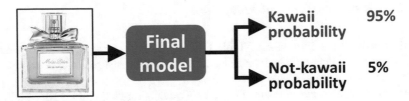

Fig. 12.4 Example of using final model to evaluate kawaiiness of cosmetic bottle images

12.3.5 Model Construction

Since the classification results showed a certain consistency, we constructed two final models for the Thai and Japanese datasets. For each, a complete dataset, including all the subsets (v1 to v8), was used as a training set. Finally, we obtained a final model for the Thai participants, and another final model for the Japanese participants.

These final schemes are classification models of the kawaiiness of cosmetic bottles. Using a cosmetic bottle image as input, the model output is the probabilities of classification into kawaii and not-kawaii groups (Fig. 12.4). The kawaii probability will be used in the next step to evaluate the kawaiiness of cosmetic bottles.

12.4 Evaluation of Attributes for Kawaiiness of Cosmetic Bottles Using Model

The Deep CNN Model is typically known as a black-box function, which means that it has difficulty tracing back which attributes are effective. However, interpreting effective attributes is critical for product designers and manufacturers. Therefore, we developed a new method to evaluate attributes using our model. In this section, we describe the evaluation procedure and results.

12.4.1 Evaluation Procedure

The following is the procedure for evaluating the kawaiiness of the cosmetic bottle images:

1. Selection of images: The images that all the participants agreed were either kawaii or not-kawaii were used for the evaluation.
2. Attribute observation: We observed the tendency of the kawaii attributes for the images collected from Step 1 and made inferences about which attributes are likely to be effective for kawaiiness.

3. Evaluation of attributes using models: We modified the original images based on our inferences and applied the final models to evaluate the kawaii probabilities of both the original and modified images. Then we calculated the difference of the kawaii probabilities between the original and modified images to confirm our inferences.

12.4.2 Evaluation Results

12.4.2.1 Results of Selection of Images

For the Thai dataset, we collected the cosmetic bottle images that all eight participants identified as kawaii (43 images) and those that were not-kawaii (40 images) from the data collection results (Sect. 12.2). Similarly, for the Japanese dataset, we collected cosmetic bottle images that all 11 participants agreed were kawaii (10 images) and those that were not-kawaii (29 images).

12.4.2.2 Results of Attribute Observation

For both the results of the Thai (Table 12.5) and Japanese (Table 12.6) datasets, there were a variety of cap ornamentation of the images in the kawaii groups. On the contrary, there were generally no objects of cap ornamentation of those in the not-kawaii group. Therefore, we inferred that kawaii caps are effective attributes to increase kawaiiness.

Table 12.5 Cosmetic bottle images grouped by objects on bottle caps for Thai participants

Group of cosmetic bottle image	Number of images	
	Kawaii	Not-kawaii
Ribbon	11	0
Flower	10	0
Miscellaneous objects	14	0
No objects	8	40

Table 12.6 Cosmetic bottle images grouped by objects on bottle caps for Japanese participants

Group of cosmetic bottle image	Number of images	
	Kawaii	Not-kawaii
Ribbon	4	0
Flower	1	0
Miscellaneous objects	1	2
No objects	4	25

From our previous research (Chap. 11 and [9]), such objects as ribbons and flowers were effective attributes for the kawaiiness of products. Therefore, to evaluate the effect of those objects, we divided the images into four groups based on the objects on the bottle caps: (1) ribbons, (2) flowers, (3) such miscellaneous objects as leaves, butterflies, hats, etc., and (4) no objects.

12.4.2.3 Results of Evaluation of Attributes Using Model

To test our inferences, we modified the images by removing their bottle caps and obtained two sets of images: with bottle caps (original images) and without them (modified images). Next, we separately evaluated the kawaiiness of these two sets using the model of Thai and Japanese participants.

As shown in Fig. 12.5, we used cosmetic bottle images with and without caps as the model's input. The model's output of each image was the probabilities of classification into kawaii and not-kawaii groups. We calculated the differences of the kawaii probabilities of the cosmetic bottle images with and without caps. This value indicates the effect of the bottle caps on the kawaiiness of each cosmetic bottle image.

Next, we calculated the mean differences of the kawaii probabilities from the differences of the kawaii probabilities of all the images in each group of cap ornamentation and performed independent-sample t-tests for the statistically significant mean differences between two groups.

For the Thai dataset, the results (Fig. 12.6) are described as follows:

- The mean differences of the kawaii probabilities between the not-kawaii group and each of the groups of kawaii images were significantly different where $p <$

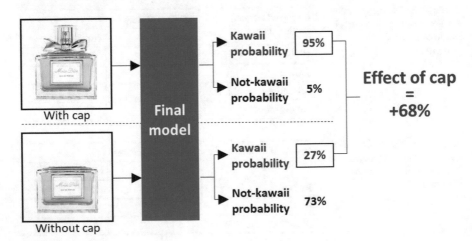

Fig. 12.5 Method to calculate differences of kawaii probabilities between cosmetic bottle images with/without caps

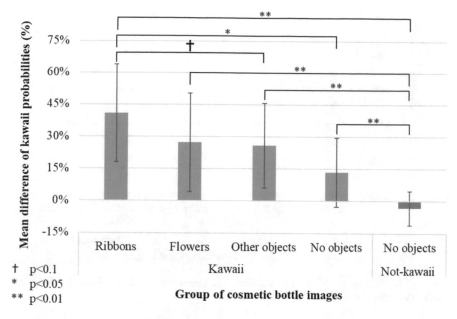

Fig. 12.6 Mean difference of kawaii probabilities among cosmetic bottle images grouped by objects on the bottle caps for Thai participants

0.01. These results indicate that the kawaii caps had a larger effect on kawaiiness than the not-kawaii caps.

- The mean differences of the kawaii probabilities between the groups with ribbons and no objects were significantly different where $p < 0.05$, and those of other objects were significantly different where $p < 0.1$. This result indicates that ribbon caps had a larger effect on kawaiiness than caps with other objects or no objects among the kawaii images.
- There were no significant differences between the groups with ribbons and flowers. This result indicates that the kawaii caps, especially ribbons and flowers, effectively increased the kawaiiness of the cosmetic bottles.

For the results of the Japanese datasets (Fig. 12.7), the mean difference of the kawaii images with ribbons was the largest (29.3%), while those of the other groups were small. However, the independent-sample t-tests did not show any statistically significant mean differences between any two groups of cosmetic bottle images. Therefore, we obtained only the tendency that ribbons are likely to increase the kawaiiness of cosmetic bottles.

Based on the above results, we conclude that the tendencies of the ribbon and flower caps are candidates for attributes to increase the kawaiiness of cosmetic bottle images.

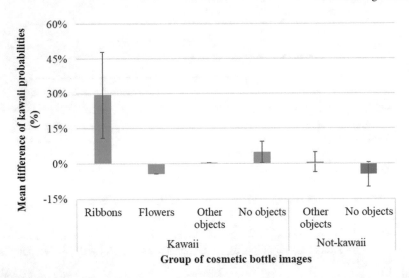

Fig. 12.7 Mean difference of kawaii probabilities among cosmetic bottle images grouped by objects on the bottle caps for Japanese participants

12.5 Evaluation of Attributes of Cosmetic Bottles Using Model and Eye Tracking

In the previous section, we identified the candidates for effective attributes for kawaii cosmetic bottles based on the predicted results by our model. In this section, we experimentally clarified them using eye tracking.

12.5.1 Observation of Attributes

To increase the number of attribute candidates, we observed the cosmetic bottle images again. We selected and observed the images that 10 or 11 Japanese participants agreed were kawaii (67 images) or not-kawaii (29 images) from the data collection result (Sect. 12.2). In this experiment, we employed the result of ten participants because the number of images from 11 participants was too few to observe the tendencies.

Our observations were designed to make inferences about the attributes of kawaii cosmetic bottles. We compared the tendency of the attributes between the kawaii and not-kawaii groups. In this research, we only focused on three hues: monochrome, blue, and pink. To maintain the overall impression of each bottle, we retained the original hues of the bottles and selected the three attribute candidates described below.

Table 12.7 Cosmetic bottle images used as candidates of cap ornamentation

Hue	Cap ornamentation			
	Flower		Ribbon	
Monochrome	F1	F2	R1	R2
Blue		F3		R3
Pink	F4	F5	R4	R5

Fig. 12.8 Two bottle shape candidates

(1) Cap Ornamentation: 11 of the images in the kawaii group had flowers and 11 also had ribbons. We selected ten candidates of cap ornamentation (Table 12.7) based on the following conditions.

- We excluded excessively large caps to maintain a size balance for further eye movement analysis.
- If the caps had a similar appearance, only one was selected.

(2) Bottle shape: The images in the kawaii group tended to be round, and those in the not-kawaii group tended to be square. Based on this tendency, we selected two different bottle shapes (Fig. 12.8).

(3) Bottle Lightness/Saturation (L/S): The images in the kawaii group tended to have transparent bottles, which were related to low L/S values. Those in the not-kawaii group tended to be black or a solid color that was related to high L/S values. Based on this tendency, we set three L/S levels (dark, bright, and brilliant) by adjusting the combination of lightness and saturation (Table 12.8).

Table 12.8 Three levels of lightness/saturation of each hue

Level	Lightness	Saturation	Hue		
			Monochrome	Blue	Pink
1 (Dark)	0	0			
2 (Bright)	-15	50			
3 (Brilliant)	-30	100			

Fig. 12.9 Examples of modified images of F1 (flower) cap ornamentation

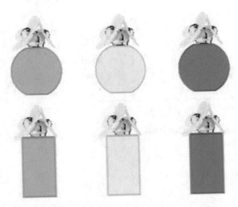

12.5.2   Candidates of Cosmetic Bottle Images

From the observation results of cap ornamentation, we only employed ten cap images. We modified their bottle shapes and their lightness/saturation based on their original hues and obtained 60 modified images: 24 monochrome, 12 blue, and 24 pink. Examples of the modified images are shown in Fig. 12.9.

12.5.3   Comparison System

We modified our previous research's comparison system in Chap. 6. Our modified system used 60 candidates of cosmetic bottle images as visual stimuli. We only compared images among identical hues. For each hue, the images were displayed in pairs. The total number of compared pairs was 60 (24 for monochrome, 12 for blue, 24 for pink), and each image just appeared twice. For each comparison pair, all three attributes were different. We arranged the 60 pairs in such a way to avoid identical hues and images between consecutive pairs.

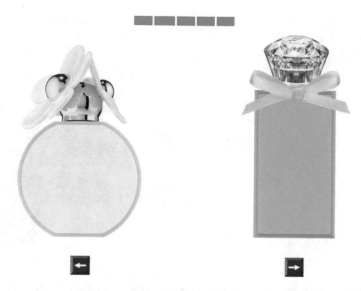

Fig. 12.10   Screenshot of comparison system displaying two cosmetic bottle images and countdown timer

The following is the structure of our comparison system:

1. The first page explained the questionnaire.
2. The consent form briefly explained the experiment and asked for permission to use their data.
3. The cosmetic bottle image selections were explained as follows:

   a. Comparison of images of cosmetic bottles.
   b. Cross sign (+) appeared in the middle of the display for 2.5 s to fix their eyes in the same position before each comparison.
   c. Pairs of cosmetic bottle images were randomly displayed with a 5-second timer (Fig. 12.10). Participants selected the more kawaii ones of 60 pairs using the keyboard's left or right arrow keys.

4. In free description questionnaires, the participants described the criteria on which they selected the kawaii cosmetic bottle images.
5. The last page explained that the comparison was finished.
6. The selection and questionnaire results were saved in a database.

12.5.4   Experimental Setup and Procedure

Figure 12.11 shows the experimental setup. The comparison system was accessed from an eye-tracking system through a web browser whose system ran on a separate

Fig. 12.11 Experimental setup where participant looked at images on PC monitor while eye-tracking system recorded eye movements

PC due to limited resources. The eye-tracking system recorded the eye movements by a TM3 nonintrusive eye tracker (EyeTech Digital Systems, Inc.) and QG-PLUS software (DITECT Co., Ltd.). We used a 19-inch LCD monitor with 1280×1024 pixel resolution.

The following is the experimental procedures:

1. The participants sat on chairs in front of a PC.
2. They read the experiment's explanation.
3. The experimenter calibrated the eyes of the participants.
4. The experimenter showed the comparison system and started recording their eye movements.
5. They selected the more kawaii cosmetic bottle images from 60 pairs.
6. They answered questionnaires.
7. The experimenter stopped recording their eye movements.

12.5.5 Experimental Results

12.5.5.1 Participants

Since females are usually more interested in cosmetic bottles than males [10], we recruited only female participants for this experiment, which was performed with 14 Japanese female volunteers, all of whom were university students in their 20s. We used for further analysis of the experimental results of ten participants whose eye-tracking data were successfully collected.

12.5.5.2   Cumulative Results

We collected the cumulative results (the kawaii scores) from the selection results of each participant. For each participant, all of the images in each hue have maximum score at 2. For the monochrome and pink images, the total scores were 24. For the blue images, the total score was 12. Then, we calculated the average scores of each image from the scores of all the participants. Finally, we normalized the scores of all the images.

Since we compared the cumulative results with results of eye-tracking data, we only analyzed the results from the ten participants whose eye-tracking data were successfully recorded. We analyzed the effects of the three attributes on each hue. The following are our results (Table 12.9):

(1) Monochrome images: We found statistically significant main effects of cap ($p < 0.05$) and shape ($p < 0.05$). However, the results did not show a statistically significant main effect of lightness/saturation or any interaction effects. The results indicated that caps and shapes were effective attributes for kawaii monochrome cosmetic bottle images.

From the results of Tukey's post-hoc tests, we identified the following statistically significant differences:

- Caps: flower > ribbon ($p < 0.05$).
- Shape: round > square ($p < 0.05$).

(2) Blue images: We found statistically significant main effects of cap ($p < 0.01$) and lightness/saturation ($p < 0.01$). However, the results did not show a statistically significant main effect of shape or any interaction effects. The results indicated that caps and lightness/saturation were effective attributes for blue kawaii cosmetic bottle images.

From the results of Tukey's post-hoc tests, we identified the following statistically significant differences:

Table 12.9 Results of three-factor ANOVA for kawaii scores of each hue

Attribute	P-value		
	Monochrome	Blue	Pink
Cap	0.040^*	0.000^{**}	0.000^{**}
Shape	0.021^*	0.197	0.029^*
L/S	0.450	0.005^{**}	0.024^*
Cap x shape	1.000	1.000	1.000
Cap x L/S	0.983	0.926	0.981
Shape x L/S	0.861	0.991	0.877
Cap x shape x L/S	0.861	0.926	0.209

** Correlation is significant at 0.01 level
* Correlation is significant at 0.05 level

- Caps: ribbon > flower (p < 0.01).
- L/S: level 2 (bright) > level 3 (brilliant) (p < 0.05).

(3) Pink images: We found statistically significant main effects of cap (p < 0.01), shape (p < 0.05), and lightness/saturation (p < 0.05). However, the results did not show any statistically significant interaction effects. They indicated that all three attributes were candidates of effective attributes for pink kawaii cosmetic bottle images.

From the results of the Tukey's post-hoc tests, we identified the following statistically significant differences:

- Caps: flower > ribbon (p < 0.01)
- Shape: round > square (p < 0.05).
- L/S: level 2 (bright) > level 3 (brilliant) (p < 0.05).

The results of Tukey's post-hoc tests between average scores for each attribute are shown in Figs. 12.12, 12.13, and 12.14.

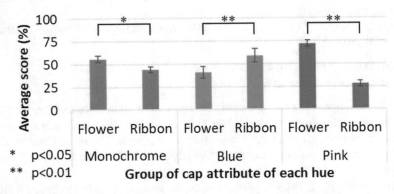

Fig. 12.12 Average scores between flower and ribbon as cap attribute of each hue

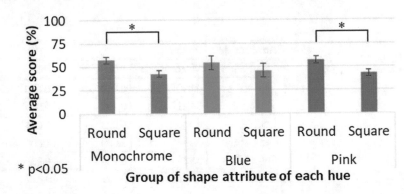

Fig. 12.13 Average scores between round and square as shape attribute of each hue

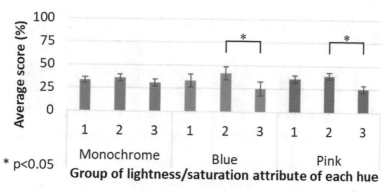

Fig. 12.14 Average scores among three levels of lightness/saturation attribute of each hue

12.5.5.3   Questionnaire Results

We summarized the questionnaire results asking about the criteria for selecting kawaii cosmetic bottle images. The keywords that the participants usually mentioned on their answers are listed below:

- Flowers and ribbons.
- Round, square shapes.
- Size of cap ornamentation.
- Color balance, color combination.

12.5.5.4   Results from Our Model

We evaluated the kawaiiness of 60 cosmetic bottle images using our model constructed in Sect. 12.3. We obtained the kawaii probabilities of the images and calculated the average kawaii probability for cap ornamentation, shape, and lightness/saturation, as shown in Figs. 12.15, 12.16, and 12.17, respectively.

From these results, we compared the tendencies with the cumulative results. The following are our results:

- For the monochrome and blue images, the tendencies of the kawaii probabilities resembled those of the cumulative results for all three attributes.
- For pink images, only the tendency of the kawaii probability for the shape attribute resembled that of the cumulative result. However, the tendencies of the other two attributes were different.

12.5.5.5   Results of Eye-Tracking Data

To analyze eye-tracking data, we employed fixation and Area of Interest (AOI). Fixation is defined as the eye state when it remains still or looks at the same spot

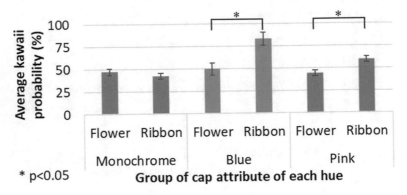

Fig. 12.15 Average kawaii probabilities between flower and ribbon as cap attribute of each hue

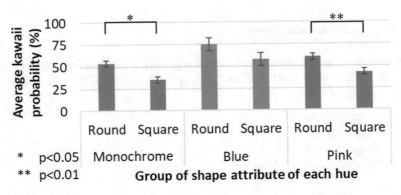

Fig. 12.16 Average kawaii probabilities between round and square as shape attribute of each hue

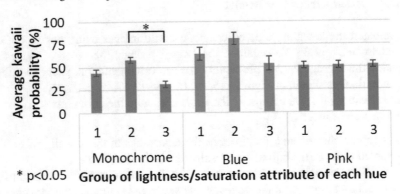

Fig. 12.17 Average kawaii probabilities among three levels of lightness/saturation attribute of each hue

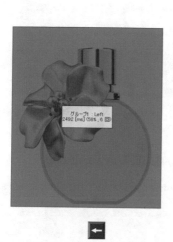

Fig. 12.18 Example of AOIs of two cosmetic bottle images whose sizes correspond to round and square shapes showing areas included in analysis of eye-tracking data

over a period of time (threshold) that we set to 200 ms. AOI is defined as the area used to include or exclude certain segments from analysis.

For this experiment's analysis, we defined two AOIs for the left-side and right-side cosmetic bottle images (Fig. 12.18) and created AOIs as squares with two different dimensions based on different widths and heights between the round and square bottle shapes. However, we kept the areas between the two AOIs to maintain the balance of the analysis areas.

We analyzed the eye-tracking data by employing three eye movement indexes from our previous research (Chap. 6). Detailed analysis is described next.

(1) Total AOI duration: We analyzed the total AOI durations between groups of each attribute. For each attribute of each hue, independent-sample t-tests were run to determine the differences in the average total AOI durations between the groups of that attribute.

The cap attribute's results, which showed statistically significant differences between the flower and ribbon caps for blue ($p < 0.01$) and pink ($p < 0.1$) cosmetic bottle images, resembled the cumulative results. We found no statistically significant differences for monochrome cosmetic bottle images. The results are shown in Fig. 12.19.

For the shape and lightness/saturation attributes, the results did not show any statistically significant differences between groups of any hues.

(2) Total number of fixations: We analyzed the total AOI durations between groups of each attribute using the same analysis method as that for the total AOI duration.

For the cap attribute, the results showed statistically significant differences between flower and ribbon caps for monochrome cosmetic bottle images ($p < 0.1$)

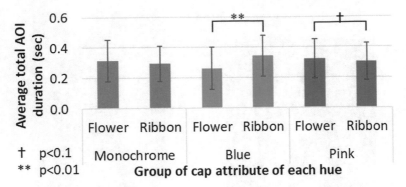

Fig. 12.19 Average total AOI durations between flower and ribbon as cap attribute of each hue

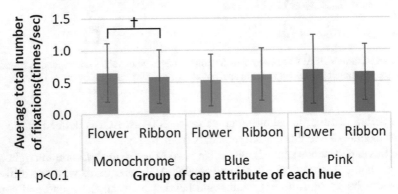

Fig. 12.20 Average total number of fixations between flower and ribbon as cap attribute of each hue

and resembled the cumulative results. We found no statistically significant differences between the blue and pink cosmetic bottle images. The results are shown in Fig. 12.20.

The lightness/saturation attribute's results, which showed statistically significant differences between levels 2 and 3 for blue cosmetic bottle images ($p < 0.1$), resembled the cumulative results. We found no statistically significant differences for the monochrome or pink cosmetic bottle images. The results are shown in Fig. 12.21.

For the shape attribute, the results did not show any statistically significant differences between groups of any hues.

(3) Number of matchings between last-eye-position images and selected images. We collected and analyzed the number of matches for each pair of comparisons between the last-eye-position image and the selected image. For each hue, an independent-sample t-test determined the differences in the average matched and unmatched numbers. The result showed significant differences in the number of matchings between the last-eye-position and the selected images ($p < 0.01$) for all hues (Fig. 12.22).

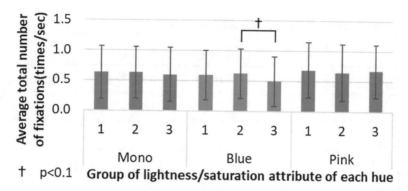

Fig. 12.21   Average total AOI durations among three levels of lightness/saturation attribute of each hue

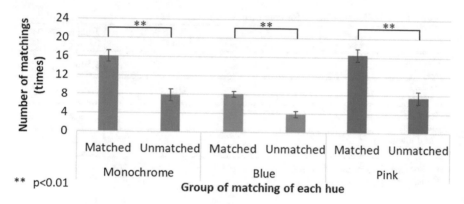

Fig. 12.22   Matchings between last-eye-position and selected images of each hue

12.5.5.6   Correlation Analysis

We performed a correlation analysis among four results: cumulative results (i.e., kawaii scores), physical attributes (i.e., caps, shapes, and lightness/saturation), two eye movement indexes (i.e., total AOI duration and total number of fixations), and the results predicted by our model (i.e., kawaii probability).

We used a Pearson product–moment correlation. Pearson's correlation coefficient (r) measures the strength of the association and the direction of the linear relationship between two variables. r ranges from -1 (perfect negative association) to $+1$ (perfect positive association). The results for each hue are described below.

(1) Monochrome images: The results (Table 12.10) indicate the following correlations:

- Moderate correlation between kawaii scores and kawaii probability ($r = 0.498$, $p < 0.01$).
- Moderate correlation between kawaii scores and total AOI duration ($r = 0.498$, $p < 0.01$).

Table 12.10 Correlation analysis for monochrome images

	Average scores	Kawaii probability	Total AOI duration	Total number of fixations
Average scores	–	0.498**	0.498**	0.055
Kawaii probability		–	0.265	-0.019
Total AOI duration			–	0.424*
Total number of fixations				–

** Correlation is significant at 0.01 level
* Correlation is significant at 0.05 level

Table 12.11 Correlation analysis for blue images

	Average scores	Kawaii probability	Total AOI duration	Total number of fixations
Average scores	–	0.900**	0.714**	0.576*
Kawaii probability		–	0.815**	0.437
Total AOI duration			–	0.680**
Total number of fixations				–

** Correlation is significant at 0.01 level
* Correlation is significant at 0.05 level

(2) **Blue images**: The results (Table 12.11) indicate the following correlations:

- Strong correlation between kawaii scores and kawaii probability ($r = 0.900$, $p < 0.01$).
- Strong correlation between kawaii scores and total AOI duration ($r = 0.714$, $p < 0.01$).
- Strong correlation between kawaii scores and total number of fixations ($r = 0.576$, $p < 0.05$).
- Strong correlation between kawaii probability and total AOI duration ($r = 0.815$, $p < 0.01$).
- Moderate correlation between kawaii probability and total number of fixations ($r = 0.437$, $p < 0.05$).

(3) **Pink images**: The results (Table 12.12) only indicate strong correlation between the kawaii scores and the total AOI duration ($r = 0.512$, $p < 0.01$). However, the negative correlation between the kawaii scores and the kawaii probability was

Table 12.12 Correlation analysis for pink images

	Average scores	Kawaii probability	Total AOI duration	Total number of fixations
Average scores	–	-0.208	0.512**	0.251
Kawaii probability		–	-0.042	0.223
Total AOI duration			–	0.144
Total number of fixations				–

** Correlation is significant at 0.01 level

contrary to our expectation, which might have been caused by the strong effects of all three attributes.

From the correlation analysis results, we obtained the following findings:

- For all the hues, the relationships between the average scores and the total AOI duration were confirmed, indicating the effectiveness of this index to measure kawaii feelings.
- For the monochrome and blue images:

  - The average scores and kawaii probabilities had similar results. However, the tendency of the pink images was surprising, perhaps caused by the effects of all three attributes. This result indicates that pink's effect is more complicated than the other hues.
  - Total AOI duration and the number of fixations had similar results.

- For blue images:

  - The relationship between the average scores and the number of fixations was confirmed, indicating the effectiveness of this index to measure kawaii feelings.
  - We confirmed the relationships among the kawaii probabilities and the two eye movement indexes, which ensured the effectiveness of the model prediction.

12.6   Conclusion

This research studied kawaii feelings for cosmetic bottles. In Sect. 12.2, we experimentally evaluated the kawaiiness of cosmetic bottle images. From our experimental results, we used the results of 8 Thai and 11 Japanese participants for model construction.

In Sect. 12.3, we constructed a model of kawaii feelings for cosmetic bottles using the Deep Convolutional Neural Network (CNN) algorithm. Our model classifies cosmetic bottle images by a binary definition of kawaiiness (kawaii or not-kawaii).

Since our Deep CNN model cannot generate candidates of effective attributes as the SVM model did, in Sect. 12.4, we developed a new method to obtain these attributes by modifying images based on particular attributes and using our model to predict the kawaii probabilities. The evaluation results showed that caps, especially with ribbons and flowers, effectively increased the kawaiiness of cosmetic bottles. With these results, we confirmed that we solved the limitations of the Deep CNN algorithm. In addition, we confirmed that our model effectively evaluated the attributes of cosmetic bottles, showing that it can evaluate not only total impressions but also particular attributes.

In Sect. 12.5, we experimentally evaluated cosmetic bottle images based on their attributes using eye tracking. We obtained the following conclusions: cap ornamentations and lightness/saturation are effective attributes, as clarified by the total AOI duration and the total number of fixations. These two attributes can be recommended to manufacturers as critical factors to increase the kawaiiness of cosmetic bottles. In addition, from the correlation analysis results, we identified strong correlations between kawaii scores and kawaii probabilities for the monochrome and blue images. These results confirmed the effectiveness of predictions using our model, suggesting its usefulness for evaluating the kawaiiness of at least monochrome and blue cosmetic bottles.

Future work will improve our model's accuracy by increasing the number of participants. We will also consider such demographic information of participants as age to construct a new model. In addition, we only clarified that cap ornamentation was an effective attribute. However, since the results were not consistent for every hue, future work will clarify other hues and other attributes.

Acknowledgements Part of this work was supported by Grant-in-Aid Scientific Research Number 26280104. We thank the students of Shibaura Institute of Technology and King Mongkut's University of Technology Thonburi for their participation.

References

1. Laohakangvalvit, T., Achalakul, T., & Ohkura, M. (2018). A Model of Kawaii Feelings for Cosmetic Bottles. In *Proceedings of the 5th International Conference on Business and Industrial Research (ICBIR)*, Bangkok.
2. Laohakangvalvit, T., Achalakul, T., & Ohkura, M. (2018). Evaluation of attributes for cosmetic bottles using eye tracking. In *Proceedings of the 20th Congress of the International Ergonomics Association (IEA)*, Florence.
3. Laohakangvalvit, T., Achalakul, T., & Ohkura, M. A Method to Obtain Effective Attributes for Attractive Cosmetic Bottles by Deep Learning. (Under Submission).
4. Azeema, N. (2016) Factors influencing the purchase decision of perfumes with habit as a mediating variable: an empirical study in Malaysia. *Indian Journal of Marketing, 46*(7).
5. Bhatt, K. (2017). A study on consumer buying behavior towards cosmetic products. *International Journal of Engineering Technology Science and Research, 4*(12), 1244–1249.
6. LeCun, Y., Bengio, T., & Hinton, G. (2015). Deep learning. *Nature, 521*, 436–444.
7. Chen, X., & Lin, X. (2014). Big data deep learning: Challenges and perspectives. *IEEE Access, 2*, 514–525.

8. Hsu, C., Change, C., & Lin, C. (2003). A practical guide to support vector classification. Technical report, Department of Computer Science, National Taiwan University.
9. Ohkura, M., Komatsu, T., Tivatansakul, S., & Saromporn, C. (2012). Comparison of evaluation of kawaii ribbons between genders and generation of Japanese. In *Proceedings of the 2012 IEEE/SICE International Symposium on System Integration (SII)*, Fukuoka.
10. McIntyre, M. P. (2013). Perfume packaging, seduction and gender. *Journal of Current Cultural Research, 5*(19), 291–311.

Chapter 13
Kawaii Perception of Artifacts Between Chinese and Japanese Cultures

Nan Qie, Pei-Luen Patrick Rau, Michiko Ohkura and Chien-Wen Tung

Abstract Chinese and Japanese culture have long been influencing each other. With the popularity of Japanese animation, comics, and game industry, Japanese kawaii culture has swept across China, especially among young Chinese people. In this chapter, we will introduce a study we conducted between Chinese and Japanese people to learn about their perceptions of kawaii designs in artifacts. This study investigates what features make an artifact kawaii, how gender and age affect kawaii perception of artifacts, and whether Chinese people perceive kawaii in artifacts in the same manner as Japanese people do. An experiment which involved both elderly and young participants from Beijing and Tokyo was conducted. Three types of artifacts were presented to the participants, and their attitudes toward different designs were recorded in the form of a questionnaire. The results indicate that culture, gender, and age can affect kawaii perception. Chinese participants and elderly Japanese participants associated kawaii closely with practical aspects. The results also show that a simple combination of kawaii elements does not necessarily contribute to the kawaii design of an artifact. Kawaii is an integrated concept and cannot be simply defined by a series of discrete elements or features.

Keywords Kawaii · Affective design · Artifacts

13.1 Kawaii Outside Japan

Kawaii has been an icon of Japanese pop culture, while the influence of kawaii is not confined to Japan. Japanese kawaii has garnered attention and interests from all over the world, resulting in considerable economic benefits. Thai teenagers and young girls are greatly affected by Japanese kawaii fashion. In 2009, a beauty contest themed as Kawaii Festa was held in Bangkok [1]. Malay researchers have concerned

N. Qie · P.-L. P. Rau (✉) · C.-W. Tung
Tsinghua University, Beijing, China
e-mail: rpl@mail.tsinghua.edu.cn

M. Ohkura
Shibaura Institute of Technology, Tokyo, Japan

© Springer Nature Singapore Pte Ltd. 2019
M. Ohkura (ed.), *Kawaii Engineering*, Springer Series on Cultural
Computing, https://doi.org/10.1007/978-981-13-7964-2_13

kawaii to be an example of a trend transition from Japan to Malaysia [2]. Taiwan has a large community of Harizu (the Mandarin term for Japanification) [3]. There was even an attempt to take advantage of kawaii in Taiwanese politics in order to construct a collective imagination and national identity [4]. The influence of kawaii is not confined to Asia. Kawaii design methods have also drawn Australian researchers' attention [5].

Several popular Japanese kawaii images have prevailed around the world. The small cat with a blank face named Hello Kitty led Pink Globalization [6]. Pokemon started as a game on Nintendo's Game Boy; quickly spread into the fields of television shows, films, toys, ancillaries, and clothes; and dominated children's consumption from approximately 1996 to 2001 [7]. Pokemon regained much attention in 2016 due to the high popularity of the location-based augmented reality game Pokemon Go. Pokemon Go quickly became the most used and profitable mobile application in 2016 with more than 500 million downloads worldwide. With the increase in popularity of these images and Japanese animation, comics, and games (ACG), kawaii becomes a representative of Japanese soft power.

13.2 Factors Considered in This Study

The standard for kawaii depends on the type of an artifact. Kawaii design is most commonly observed in stationery, toys, and small daily necessities that may have decorative properties. In addition to daily necessities, artifacts not decorated and not regarded as emotional are also considered. In this study, medical devices and high-tech products are chosen. Medical devices are usually designed to be cold and serious, and high-tech products are usually designed to be cool and apathetic. Kawaii may be a good approach to adding value to these products.

Factors including culture, gender, and age can also affect kawaii perception of artifacts. In this study, kawaii perceptions of Chinese and Japanese people were compared. China and Japan are geographically close and share East Asian cultural values. Japanese cartoons and manga now fuel a huge market worth $14.6 billion in China. Nearly all universities and high schools in Beijing have their own anime clubs. Sometimes, only hours after an anime series airs in Japan, Chinese fans are able to download a translated version from the Internet [8]. Therefore, kawaii along with Japanese ACG, is becoming very popular with the younger generation in China in recent years [9]. Despite being similar in cultural background, Chinese and Japanese aesthetics are not the same. In addition, compared with the younger generation, Chinese elderly people are not that familiar with Japanese culture. Based on the previous discussion, three research questions are considered:

RQ1: What features make an artifact kawaii?

RQ2: How do gender and age affect kawaii perceptions in artifacts?

RQ3: Do Chinese people perceive kawaii in artifacts in the same manner as Japanese people do?

In order to answer the three questions, an experiment was conducted. Through a questionnaire, Chinese and Japanese participants were asked to rate designs on how much they liked them, how willing they were to use them, and whether the designs were kawaii.

13.3   The Kawaii Perception Experiment

The experiment involved 60 Chinese and 60 Japanese participants, in either culture including 30 young (Chinese age M = 23.6, SD = 1.67; Japanese age M = 22.0, SD = 1.00) and 30 elderly people (Chinese age M = 72.3, SD = 8.09; Japanese age M = 70.3, SD = 4.72). The participants were strictly balanced in gender. The young participants were university students from Beijing and Tokyo. The Chinese elderly people were recruited from a university for the elderly and an activity center for the elderly in a university in Beijing, and the Japanese elderly people were recruited from a part-time job center for the elderly in Tokyo.

Three types of artifacts were selected as the materials: daily necessities, medical devices, and high-tech products. A product representative was chosen for each artifact type. A spoon is commonly used as a daily necessity by both the young and the elderly, and by both men and women in everyday life. In this experiment, spoons were described as used exclusively for stirring in order to unify evaluation standard. The designs of most medical devices are usually insipid and similar, which may lead to boredom during long-term usage. It is interesting to observe whether kawaii can cause any change in medical device design and how our participants perceive it. In this experiment, the blood pressure monitor was selected as the representative medical device. The blood pressure monitor is a commonly used medical device, which is especially popular among the elderly people. As for the high-tech product, a robot was used in this experiment. Robots, as the representative high-tech product, are usually expected to have a cool and mechanical design; however, robots are different from other high-tech products in that they are sometimes designed to imitate human beings and animals. Kawaii is commonly accepted as a feature in living creatures. Thus, it is conceivable to include kawaii in robot design. In this study, the robots were further qualified as family care robots with human appearance. Each product has four different designs, shown in Figs. 13.1, 13.2 and 13.3. Real spoons were used while pictures of blood pressure monitors and robots were presented on a display to the participants. The distance between the participant and the screen was controlled the same for each participant.

Most of the time, participants took part in the experiment one by one, while, at times, two elderly Japanese participants did it together owing to a limitation of time. At the beginning, the participant was informed of the general process of the experiment and was asked to sign an informed consent. After a short introduction to the spoons, four designs of spoons were presented to the participant one by one, followed by the blood pressure monitors, and finally the robots. Each time the participant saw a product, he/she was asked to rate it with respect to the following: "How much do

Fig. 13.1 Spoons used in the experiment

Fig. 13.2 Pictures of blood pressure monitors used in the experiment

Fig. 13.3 Pictures of robots used in the experiment

you like it?" (like scale), "How much do you want to use it?" (use scale), and "How kawaii do you think it is?" (kawaii scale) from 1 (not at all) to 7 (very much). The orders of the designs within each product group were random. After seeing all the products, the participants were again presented with all the four spoons, then all the four blood pressure monitors, and finally all the four robots. They were asked about their most liked and most disliked design for each product, and about why they made the choices. The whole procedure lasted about 40 min for each participant.

Mixed-ANOVA models were used to determine the main and interaction effects of culture, design, age, and gender on the like, use, and kawaii ratings. The within-subject variable was the design of the product, and the other variables were between-subject variables. Only two-way or three-way interaction effects were considered. Mauchly's test was used to test the sphericity assumption for the within-subject variable, and the degrees of freedom were corrected based on Huynh–Feldt estimates of sphericity when the sphericity assumption was violated. Paired t-tests with a Bonferroni adjustment were used for the post hoc analysis.

13.4 Results

13.4.1 Daily Necessities

Like As shown in Table 13.1 and Fig. 13.4, the effect of gender; design; the two-way interaction of age and design; the two-way interaction of culture and design; the three-way interaction of age, gender, and culture; and the three-way interaction of age, gender, and design were significant. Female participants (M = 4.87, SD = 1.42) rated significantly higher than male participants (M = 4.53, SD = 1.46) on the like scale. No significant differences were observed in the ratings of the four designs by Chinese participants (p > 0.170). In the case of Japanese participants, Design 3 was liked significantly more than the other three designs (ps < 0.016). In the case of Chinese participants, elderly female participants (M = 5.07, SD = 1.67) liked the designs significantly more than elderly male participants (M = 4.18, SD = 1.60, p = 0.004), while the ratings given by young participants were not significantly different between genders (p = 0.780). In the case of Japanese participants, young female participants (M = 5.20, SD = 0.97) liked the designs significantly more than young male participants (M = 4.48, SD = 1.51, p = 0.003), while the ratings given by elderly participants were not significantly different between genders (p = 0.220). In the case of young participants, female participants (M = 4.77, SD = 1.38) liked Design 2 significantly more than male participants (M = 3.80, SD = 1.40, p = 0.009). In the case of elderly participants, female participants (M = 5.37, SD = 1.50) liked Design 3 significantly more than male participants (M = 4.43, SD = 1.76, p = 0.031).

Use As shown in Table 13.1 and Fig. 13.5, the effect of gender; design; the two-way interaction of age and culture; the three-way interaction of age, gender, and culture; and the three-way interaction of age, gender, and design were significant. Female participants (M = 4.30, SD = 1.62) rated significantly higher than male participants (M = 3.90, SD = 1.72) on the use scale. Design 4 was rated significantly lower in usefulness than the other three designs (ps < 0.032). In the case of Chinese participants, elderly female participants (M = 4.27, SD = 1.93) rated the designs significantly more useful than the elderly male participants (M = 3.25, SD = 1.82, p = 0.004), while the ratings given by young participants were not significantly different between genders (p = 0.272). In the case of Japanese participants, young female participants (M = 4.50, SD = 1.36) rated the designs as significantly more useful than young male participants (M = 3.67, SD = 1.64, p = 0.003), while the ratings given by elderly participants were not significantly different between genders (p = 0.068). In the case of young participants, female participants (M = 4.63, SD = 1.45) rated Design 2 as significantly higher than male participants (M = 3.67,

Table 13.1 ANOVA summary table of main and interaction effects on spoon ratings

Factor	Like					Use					Kawaii				
	df_1	df_2	F	p	η^2	df_1	df_2	F	p	η^2	df_1	df_2	F	p	η^2
Age	1	112	0.77	0.382	<0.01	1	112	0.06	0.799	<0.01	1	112	0.78	0.378	<0.01
Gender	1	112	5.48	0.021[a]	0.02	1	112	6.02	0.016[a]	0.02	1	112	4.05	0.046[a]	0.01
Culture	1	112	0.41	0.521	<0.01	1	112	3.02	0.085	<0.01	1	112	3.24	0.075	0.01
Design	3	336	8.09	<0.001[b]	0.05	3	336	6.36	<0.001[b]	0.04	3	336	9.70	<0.001[b]	0.05
Age * gender	1	112	0.17	0.683	<0.01	1	112	1.04	0.309	<0.01	1	112	0.99	0.322	<0.01
Age * culture	1	112	0.22	0.641	<0.01	1	112	4.61	0.034[a]	0.01	1	112	0.32	0.574	<0.01
Gender * culture	1	112	0.99	0.322	<0.01	1	112	2.51	0.116	<0.01	1	112	0.02	0.893	<0.01
Age * design	3	336	7.03	<0.001[b]	0.04	3	336	1.00	0.394	<0.01	3	336	9.08	<0.001[b]	0.05
Gender * design	3	336	0.55	0.648	<0.01	3	336	0.99	0.398	<0.01	3	336	0.94	0.420	<0.01
Culture * design	3	336	2.87	0.037[a]	0.02	3	336	1.63	0.183	<0.01	3	336	3.16	0.025[a]	0.02
Age * gender * culture	1	112	10.74	0.001[a]	0.03	1	112	10.36	0.002[a]	0.03	1	112	2.86	0.094	<0.01
Age * gender * design	3	336	3.43	0.017[a]	0.02	3	336	4.35	0.005[a]	0.03	3	336	2.27	0.080	0.01
Age * culture * design	3	336	0.91	0.435	<0.01	3	336	2.02	0.111	<0.01	3	336	0.33	0.807	<0.01
Gender * culture * design	3	336	0.38	0.769	<0.01	3	336	1.09	0.355	<0.01	3	336	1.20	0.312	<0.01

[a] $p < 0.05$.
[b] $p < 0.001$

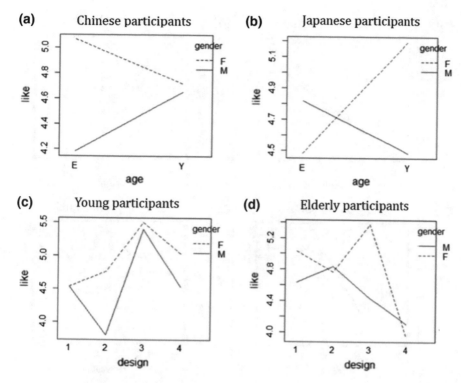

Fig. 13.4 Interaction plots of age, gender, and culture like ratings of spoons (**a–b**) and the interaction plots of age, gender, and design on like ratings of daily necessities (**c–d**)

SD = 1.42, p = 0.012), and female participants (M = 4.23, SD = 1.55) rated Design 4 significantly higher than male participants (M = 3.33, SD = 1.63, p = 0.032). In the case of elderly participants, female participants (M = 4.90, SD = 1.73) liked Design 3 significantly more than male participants (M = 3.77, SD = 1.91, p = 0.019).

Kawaii As shown in Table 13.1, the effect of gender, design, the two-way interaction of culture and design, and the two-way interaction of age and design were significant. Female participants (M = 5.23, SD = 1.46) rated significantly higher than male participants (M = 4.92, SD = 1.57) on the kawaii scale. Japanese participants rated Design 3 as significantly more kawaii than Design 1 or Design 2 (ps < 0.001), while there was no significant difference in the ratings provided by Chinese participants. Young participants rated Design 4 and Design 3 as significantly more kawaii than Design 2 or Design 1 (ps < 0.001), while elderly participants rated Design 3 as significantly more kawaii than design 4 (p = 0.033).

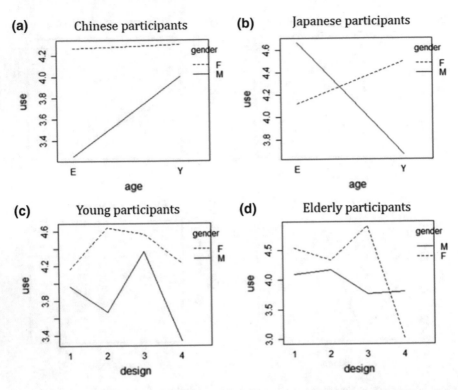

Fig. 13.5 Interaction plots of age, gender, and culture on use ratings of spoons (**a–b**) and the interaction plots of age, gender, and design on use ratings of daily necessities (**c–d**)

13.4.2 Medical Devices

Like As shown in Table 13.2, the effect of age, the two-way interaction of age and culture, and the two-way interaction of gender and design were significant. Elderly Chinese participants (M = 4.60, SD = 1.19) liked the designs significantly more than young participants (M = 3.93, SD = 1.35, p < 0.001), while no significant difference was observed in the ratings of Japanese young and elderly participants (p = 0.869). Male participants (M = 4.90, SD = 1.31) liked Design 1 significantly more than female participants (M = 4.45, SD = 1.08, p = 0.042). Female participants (M = 4.68, SD = 1.10) liked Design 2 significantly more than male participants (M = 4.12, SD = 1.46, p = 0.018), and female participants (M = 4.57, SD = 1.43) liked Design 4 significantly more than male participants (M = 4.02, SD = 1.58, p = 0.048).

Use As shown in Table 13.2, the effect of age, design, the two-way interaction of age and culture, and the two-way interaction of gender and design were significant. Chinese elderly participants (M = 4.78, SD = 1.58) rated significantly higher in usefulness than young participants (M = 4.01, SD = 1.45, p < 0.001), while no significantly different ratings were observed between Japanese young and elderly

Table 13.2 ANOVA summary table of main and interaction effects on blood pressure monitor ratings

Factor	Like					Use					Kawaii				
	df_1	df_2	F	p	η^2	df_1	df_2	F	p	η^2	df_1	df_2	F	p	η^2
Age	1	112	9.62	0.002[a]	0.03	0.9	105.3	5.18	0.025[a]	0.02	1	112	11.11	0.001[a]	0.03
Gender	1	112	2.59	0.111	<0.01	0.9	105.3	2.54	0.114	<0.01	1	112	1.15	0.285	<0.01
Culture	1	112	2.23	0.138	<0.01	0.9	105.3	0.39	0.533	<0.01	1	112	1.04	0.310	<0.01
Design	3	336	2.40	0.068	0.01	2.8	315.8	4.75	0.003[a]	0.03	3	336	15.28	<0.001[b]	0.08
Age * gender	1	112	0.40	0.529	<0.01	0.9	105.3	0.03	0.865	<0.01	1	112	0.27	0.602	<0.01
Age * culture	1	112	80.58	0.004[a]	0.03	0.9	105.3	8.76	0.004[a]	0.03	1	112	7.74	0.006[a]	0.02
Gender * culture	1	112	<0.01	0.954	<0.01	0.9	105.3	0.05	0.820	<0.01	1	112	0.64	0.426	0.03
Age * design	3	336	0.24	0.869	<0.01	2.8	315.8	0.93	0.427	<0.01	3	336	4.25	0.006[a]	0.06
Gender * design	3	336	4.80	0.003[a]	0.03	2.8	315.8	6.16	<0.001[b]	0.03	3	336	5.38	0.001[a]	<0.01
Culture * design	3	336	1.07	0.361	<0.01	2.8	315.8	0.76	0.516	<0.01	3	336	11.25	<0.001[b]	<0.01
Age * gender * culture	1	112	1.32	0.253	<0.01	0.9	105.3	0.94	0.335	<0.01	1	112	0.06	0.805	<0.01
Age * gender * design	3	336	0.86	0.463	<0.01	2.8	315.8	1.19	0.313	<0.01	3	336	1.51	0.211	<0.01
Age * culture * design	3	336	2.22	0.085	0.01	2.8	315.8	0.54	0.657	<0.01	3	336	1.14	0.700	<0.01
Gender * culture * design	3	336	0.36	0.785	<0.01	2.8	315.8	0.23	0.876	<0.01	3	336	0.47	0.396	<0.01

[a] $p < 0.05$.
[b] $p < 0.001$

participants (p = 0.499). Male participants (M = 4.97, SD = 1.10) rated Design 1 as significantly higher in usefulness than female participants (M = 4.43, SD = 1.14, p = 0.010). Female participants (M = 4.68, SD = 1.05) rated Design 2 significantly higher in usefulness than male participants (M = 4.03, SD = 1.54, p = 0.008), and female participants (M = 4.45, SD = 1.49) rated Design 4 significantly higher in usefulness than male participants (M = 3.83, SD = 1.54, p = 0.028).

Kawaii As shown in Table 13.2, the effect of age; design; the two-way interaction of age and culture; the two-way interaction of age and design; the two-way interaction of gender and design; and the two-way interaction of culture and design were significant. Chinese elderly participants (M = 4.75, SD = 1.64) rated significantly higher in kawaii than young participants (M = 3.83, SD = 1.49, p < 0.001), while no significantly different ratings were observed between Japanese young and elderly participants (p = 0.663). Elderly participants (M = 4.25, SD = 1.62) rated Design 1 as significantly more kawaii than young participants (M = 3.08, SD = 1.24, p < 0.001), and elderly participants (M = 4.87, SD = 1.32) rated Design 3 as significantly more kawaii than young participants (M = 4.35, SD = 1.45, p = 0.043). Male participants (M = 3.98, SD = 1.61) rated Design 1 as significantly more kawaii than female participants (M = 3.35, SD = 1.44, p = 0.025), while female participants (M = 4.92, SD = 1.51) rated Design 4 as significantly more kawaii than male participants (M = 4.28, SD = 1.64, p = 0.029). Chinese participants (M = 4.18, SD = 1.62) rated Design 1 as significantly more kawaii than Japanese participants (M = 3.15, SD = 1.30, p < 0.001). Japanese participants (M = 4.92, SD = 1.14) rated Design 3 as significantly more kawaii than Chinese participants (M = 4.30, SD = 1.58, p = 0.016), and Japanese participants (M = 4.92, SD = 1.29) rated Design 4 as significantly more kawaii than Chinese participants (M = 4.28, SD = 1.81, p = 0.030).

13.4.3 High-Tech Products

Like As shown in Table 13.3, the effect of age, design, and the two-way interaction of gender and design were significant. Elderly participants (M = 4.93, SD = 1.36) liked the designs significantly more than young participants (M = 4.38, SD = 1.41, p < 0.001). Male participants (M = 5.47, SD = 1.10) liked Design 1 significantly more than female participants (M = 4.87, SD = 1.35, p = 0.009).

Use As shown in Table 13.3, the effect of age, design, the two-way interaction of age and culture, and the two-way interaction of gender and design were significant. Chinese elderly participants (M = 4.82, SD = 1.51) rated significantly higher in usefulness than young participants did (M = 4.48, SD = 1.60, p < 0.001), while no significantly different ratings were observed between Japanese young and the elderly participants (p = 0.501). Female participants (M = 4.77, SD = 1.33) rated Design 2 significantly higher in usefulness than male participants did (M = 4.48, SD = 1.28, p = 0.025).

Table 13.3 ANOVA summary table of main and interaction effects on robot ratings

Factor	Like					Use					Kawaii				
	df_1	df_2	F	p	η^2	df_1	df_2	F	p	η^2	df_1	df_2	F	p	η^2
Age	0.9	106.0	12.70	<0.001[b]	0.04	0.9	106.1	16.26	<0.001[b]	0.06	1	112	21.72	<0.001[b]	0.07
Gender	0.9	106.0	0.35	0.554	<0.01	0.9	106.1	0.31	0.580	<0.01	1	112	0.08	0.782	<0.01
Culture	0.9	106.0	0.19	0.666	<0.01	0.9	106.1	2.46	0.119	<0.01	1	112	0.01	0.940	<0.01
Design	2.8	317.9	11.62	<0.001[b]	0.06	2.8	318.3	11.49	<0.001[b]	0.05	3	336	2.17	0.091	0.01
Age * gender	0.9	106.0	0.14	0.706	<0.01	0.9	106.1	0.21	0.647	<0.01	1	112	0.02	0.900	<0.01
Age * culture	0.9	106.0	1.41	0.237	<0.01	0.9	106.1	4.21	0.042[a]	<0.01	1	112	2.52	0.115	<0.01
Gender * culture	0.9	106.0	0.84	0.361	<0.01	0.9	106.1	0.21	0.647	0.02	1	112	0.18	0.669	<0.01
Age * design	2.8	317.9	0.76	0.515	<0.01	2.8	318.3	1.87	0.135	<0.01	3	336	2.91	0.035[a]	0.02
Gender * design	2.8	317.9	3.65	0.013[a]	0.02	2.8	318.3	3.82	0.010[a]	<0.01	3	336	1.32	0.267	<0.01
Culture * design	2.8	317.9	0.76	0.515	<0.01	2.8	318.3	0.84	0.471	<0.01	3	336	0.71	0.547	<0.01
Age * gender * culture	0.9	106.0	0.49	0.484	<0.01	0.9	106.1	0.98	0.324	<0.01	1	112	0.11	0.744	<0.01
Age * gender * design	2.8	317.9	0.46	0.713	<0.01	2.8	318.3	0.07	0.976	<0.01	3	336	0.48	0.698	<0.01
Age * culture * design	2.8	317.9	0.91	0.439	<0.01	2.8	318.3	1.68	0.171	<0.01	3	336	4.45	0.004[a]	0.02
Gender * culture * design	2.8	317.9	0.20	0.899	<0.01	2.8	318.3	0.42	0.740	<0.01	3	336	0.50	0.734	<0.01

[a] $p < 0.05$.
[b] $p < 0.001$

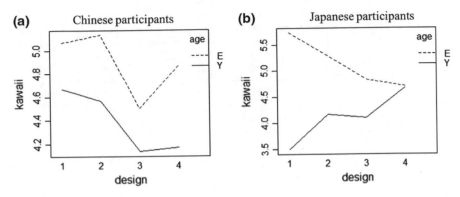

Fig. 13.6 Interaction plots of age, design, and culture on kawaii ratings of high-tech product (**a–b**)

Kawaii As shown in Table 13.3 and Fig. 13.6, the effect of age, the two-way interaction of age and design, and the three-way interaction of age, culture, and design were significant. There was no significant difference in the ratings of the designs in the cases of Chinese young and elderly participants (ps > 0.729), while Japanese young and elderly participants had opposite opinions of the four robot designs. Japanese young participants rated Design 4 (M = 4.67, SD = 1.37) as significantly more kawaii than Design 1 (M = 3.50, SD = 1.22, p = 0.009), while Japanese elderly participants rated Design 1 (M = 5.73, SD = 4.70) as significantly more kawaii than Design 3 (M = 4.83, SD = 1.46, p = 0.042) and Design 4 (M = 4.70, SD = 1.51, p = 0.012).

13.5   What Makes an Artifact Kawaii

The ratings on the like, use, and kawaii scales were all highly correlated (ps < 0.001). The ratings of the participants on the like and use scales were almost the same, while the results on the kawaii scale sometimes showed different tendencies. The meaning of "How much do you want to use it?" is very similar to "How much do you like it?", but kawaii is very different from the other two. From the results, it is shown that the participants neither equate kawaii to "liking" or "wanting to use," nor regard kawaii as a simple aesthetic concept. When participants were asked to provide reasons for their choices of most kawaii and least kawaii designs, ease of use and decorative features of colors and textures were both mentioned. Most participants considered both practicality and aesthetics when rating on the kawaii scale.

Although previous research has found rules regarding kawaii colors, kawaii shapes, and kawaii textures for Japanese people [10–12], the results of the current research indicate that the application of these rules should depend on product type. The four designs for the blood pressure monitors and robots were pure blue, blue with dots, pure pink, and pink with dots respectively, but the participants rated them differently. In the case of blood pressure monitors, Japanese participants from both age groups thought that the pink designs (Design 3 and Design 4) were more kawaii. In the case of robots, young Japanese participants preferred the design with pink dots (Design 4) while elderly participants rated the pure blue design (Design 1) as the most kawaii. The different results obtained from the two product types may be due to the expectation of the functionality of the product. Both elderly and young participants share similar expectations regarding the functionality of the blood pressure monitors. Robots in this experiment were described as family care robots. Elderly participants may expect more with respect to functionalities, such as holding their arms to avoid slipping and helping them do physical work, while young participants may expect more with respect to functionalities, such as chatting and playing with them. When asked about the reasons, more than half of the Japanese elderly participants who chose Design 1 as the most kawaii mentioned reasons like "powerful" and "reliable." The pure blue robot (Design 1) presented a masculine and powerful image, which meant reliability for elderly participants, while the pink and dotted robot (Design 4) presented a gentle and warm image, which meant friendliness for young participants.

13.6 The Effect of Age and Gender

In general, elderly participants gave higher ratings than young participants, and female participants gave higher ratings than male participants. Female participants preferred pink designs and male participants preferred blue designs in the cases of medical devices and high-tech products.

Japanese participants had diverse opinions of what was kawaii and what was not. Japanese young participants associated kawaii more with aesthetic aspects, while elderly participants associated kawaii more with practical aspects. Spoon Design 4 had a bear figurine at the top of the spoon holder, which was cute but might obstruct its use; therefore, it was rated as the lowest on the use scale. Japanese elderly participants rated it lowest on the kawaii scale with reasons such as "hard to use and clean" or "it is designed for a grandchild," while Japanese young participants rated it as the most kawaii design. The same observation was made in the high-tech product results. On the use scale, robot Design 1 with pure blue color was rated the highest, while Design 4 with pink color and dots was rated the lowest. Japanese elderly participants thought Design 1 was the most kawaii, while Japanese young participants thought Design 4 was the most kawaii. The bear figurine, pink color, and dots were all typical elements that the participants mentioned as reasons for their choice of most kawaii design. Japanese young participants were sensitive to these kawaii elements, and they could easily tell apart kawaii and practicality. On the other side, Japanese elderly participants thought that "wanting to use it" was almost the same as "kawaii". This may be because, in Japan, young people have a better understanding of kawaii in artifacts, while elderly people tend to experience difficulty in relating the concept of kawaii to nonliving things [10].

13.7 The Effect of Culture

Both Chinese young and elderly participants rated similarly to Japanese elderly participants, and they rated on the three scales consistently. Owing to the lack of previous research on kawaii element preference among Chinese participants, there could be several possible explanations for this observation. One possible reason is that the kawaii elements identified by Japanese participants are not suitably kawaii for Chinese participants. For example, pink is considered a kawaii color in Japan, but it may not be considered kawaii in China. The ecological valance theory indicates that people perceive colors strongly associated with the corresponding objects [13]. Pink is the color of Sakura, which has a strong hint of the beautiful, soft, melancholy, and delicate images in Japanese culture; therefore, it is conceivable for Japanese people to relate pink to kawaii. This tie between pink and kawaii cannot be simply applied to Chinese culture. The second possible reason is that owing to the transmission route of kawaii culture, Chinese people understand only a specific type of kawaii. Most Chinese people become familiar with kawaii through ACG products, which involve mainly kawaii cartoon characters. Based on these characters, Chinese people may become familiar with kawaii animals, kawaii schoolgirls, kawaii hairstyles, kawaii clothing, kawaii facial expressions, and so on., but they are unfamiliar with the kawaii concept with respect to artifacts. Chinese people learn kawaii from a series of kawaii representatives, but they may not understand the emotional value of kawaii. The third possible reason is the same as that for the case of Japanese elderly participants—Chinese people learn only the original concept of kawaii, and thus, it is difficult for them to relate kawaii to nonliving things.

References

1. Toyoshima, N. (2015). Kawaii fashion in Thailand: The consumption of cuteness from Japan. *Journal of Asia-Pacific Studies (Waseda University)*, *24*, 191–210.
2. Rahman, J. A., Fudzee, F. M., Kijima, A., & Furuya, S. (2013). Re-cognition of Kawaii trend in catch-up and spread structure. *Bulletin of Japanese Society for the Science of Design*, *60*(3), 3_21–3_28.
3. Lee, M.-T. (2004). *Absorbing 'Japan': Transnational media, cross-cultural consumption, and identity practice in contemporary Taiwan*. Ph.D. dissertation, King's College/Department of Social Anthropology, University of Cambridge.
4. Chuang, Y. C. (2011). Kawaii in Taiwan politics. *IJAPS*, *7*(3).
5. Glover, J., Fennessy, L., & Varadarajan, S. (2015). Apprehending Kawaii for industrial design theory. In *Proceedings of IASDR2015*, 789–804.
6. Yano, C. R. (2013). *Pink globalization: Hello Kitty's trek across the Pacific*. Duke University Press.
7. Buckingham, D., Sefton-Green, J., Allison, A., Iwabuchi, K., & Tobin, J. (2004). *Pikachu's global adventure: The rise and fall of Pokémon*. Duke University Press.
8. Cooper-Chen, A. (2011). Japan's illustrated storytelling: A thematic analysis of globalized Anime and Manga. *Keio Communication Review, 33*, 85–98.
9. jpn199miaoyangfinalblog. (2014). Japanese pop culture in China. Retrieved from http://jpn199miaoyangfinalblog.tumblr.com/.

10. Ohkura, M., & Aoto, T. (2007). Systematic study for "Kawaii" products. In *Proceedings of the 1st International Conference on Kansei Engineering and Emotion Research 2007.*
11. Ohkura, M., Goto, S., & Aoto, T. (2009). Systematic study for'Kawaii'products: Study on Kawaii colors using virtual objects. In *Proceedings of the 13th International Conference on Human-Computer Interaction*, 633–637.
12. Ohkura, M., Konuma, A., Murai, S., & Aoto, T. (2008). Systematic study for "kawaii" products (the second report)-commpmrison of "kawaii" colors and shapes. In *Proceedings of SICE Annual 2008*, 481–484.
13. Palmer, S. E., & Schloss, K. B. (2010). An ecological valence theory of human color preference. *Proceedings of the National Academy of Sciences, 107*(19), 8877–8882.

Chapter 14
How "Kawaii" Became the Key Concept of Our Tiny Electric Car

Shinsuke Ito

Abstract RimOnO Prototype 01, a tiny electric car with a textile body, won the highest award at the 4th Kawaii Award in 2016 hosted by the Japan Society of Kansei Engineering. In this chapter, Shinsuke Ito, the CEO of RimOnO Corporation, explains how a group of people with different backgrounds teamed up to create this tiny, appealing car and how the prototype came to have its current design.

Keywords Kawaii · Electric car · RimOnO · Textile

14.1 Introduction

On September 9, 2016, our prototype electric car, RimOnO Prototype 01 (Fig. 14.1), won the highest award of the 4th Kawaii Award hosted by the Japan Society of Kansei Engineering.

Two years before, I founded the RimOnO Corporation with Kota Nezu, a former Toyota designer, to develop a tiny electric car that would be universally attractive to users regardless of gender, age, or nationality. The key concept was to make it "kawaii" or cute, so that the car would be welcomed not only by car lovers but also by those with less interest in cars.

In this chapter, I explain the birth of this project and how we ended up with a textile exterior car with very kawaii features.

S. Ito (✉)
RimOnO Corporation, Tokyo, Japan
e-mail: s-ito@rimono.jp

Fig. 14.1 "RimOnO
Prototype 01"

14.2 History of the Development of RimOnO

14.2.1 An Electric Car as a "Moving Room"

Before I founded RimOnO, I worked in the Ministry of Economy, Trade and Industry for 15 years. In 2006, I met a group of people from the Tokyo Electric Power Company (TEPCO) who were very eager to introduce electric cars to homes and offices. One of their proposals, which drew my attention, was to park an electric car inside houses or buildings and to use it as a "moving room" (Fig. 14.2). They focused on the unique features of an electric car: no combustion, no exhaust, and no noise. Because of these features, it might fit more closely into people's living environments. This proposal sparked special feelings in me for electric cars.

Fig. 14.2 Introducing
electric cars into homes and
offices

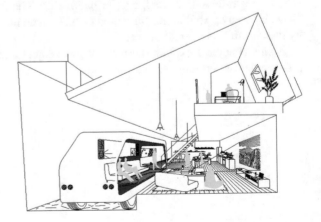

Fig. 14.3 ZecOO: an
electric motorcycle

Fig. 14.3 ZecOO: an electric motorcycle

14.2.2 Encounter with Kota Nezu and His Electric Motorcycle: ZecOO

During my time at the Ministry, I launched several projects on electric cars, including choosing eight cities to enhance the promotion of electric cars and introducing a smart grid system using them as power storage devices. However, all the electric cars used in these projects or sold in the market differed from my ideal image.

In 2012, I saw an electric motorcycle called zecOO (Fig. 14.3) on TV and was fascinated by its innovative look and features. It was designed by Kota Nezu, a former designer at Toyota, and was developed with some small and medium-sized enterprises in Japan. After I saw the video, I managed within the Ministry to find out how to contact him and met him in person.

14.3 Developing a Textile-Covered Car

After several meetings with Kota and seeing some of the vehicles he had designed, I reached 2 conclusions: A) a kawaii car might attract many users regardless of gender and age and B) a tiny car is appealing because it gives the driver and the passenger a sense of sharing a tight space. In July 2014, I left the Ministry and founded the RimOnO Corporation to develop a kawaii, tiny car with Kota.

When Kota and I first started discussing the concept of our new car, he insisted that we choose new material for it to differentiate it from conventional cars. He suggested using a waterproof textile as an exterior of our car. At first, I was very skeptical about using a textile for a car; however, his first prototype image (Fig. 14.4) excited me so much that all my concerns melted away. With this design, our new car became the cutest car ever!

Fig. 14.4 Initial image

14.4 Struggling with Design Changes

Kota started working on the CAD 3D modeling of the first prototype of our new car. However, he was concerned that a round, 3D design might easily cause wrinkles in a textile body, reducing its attractiveness. He solved this problem by coming up with a completely different design that generally consisted of quadric surfaces (Fig. 14.5).

When I first saw his second design, I was shocked because I did not think it was nearly kawaii enough. One day, I summoned the nerve to admit to him that "Actually, I don't think this car is very kawaii…" The situation started to change when the engineers of our engineering partner, Dream Design Corporation, secretly created a full-scale model of the newly designed RimOnO (Fig. 14.6). My worries evaporated when Kota began to consider redesigning the car's shape after getting inside of the model. Finally, our kawaii design was completed (Fig. 14.1).

Fig. 14.5 3D model of RimOnO's second design

Fig. 14.6 Full-scale model
of second design

14.5 Main Features of Tiny Electric Car RimOnO Prototype 01

When I developed the RimOnO Prototype 01, its most important aspect was being as kawaii as possible. Not being a car lover myself, I wanted to ensure that even people with little interest in cars would fall in love with its design. Our prototype has three primary kawaii features: (1) a front design that resembles an animation character; (2) design stitches on the textile body that suggest a teddy bear; (3) kawaii door knobs and car emblems (Fig. 14.7). All of these kawaii features were achieved not only by Kota's design genius but also thanks to our development partners who diligently worked to make an appealing car.

Another key feature of this car is its seat layout (Fig. 14.8). The back passenger seat is so close to the driver's seat that it resembles a motorcycle and creates a feeling of intimacy between the driver and the passenger. We saw big smiles on all the people who had the chance to ride in RimOnO. One person commented: "I felt like I was in a time machine that took me back to my childhood."

The car is 2.2 m long, 1.0 m wide, 1.3 m high, and weighs 320 kg. Maintaining such a small size for a two-seat car guaranteed that senior citizens and unexperienced drivers can easily drive it. Although the prototype only travels up to 20–30 km/h, its ideal top speed is 45 km/h, which gives a feeling of safety among pedestrians. Its driver's seat is rotatable to enable senior citizens to easily enter.

Fig. 14.7 Kawaii doorknobs
and car emblems

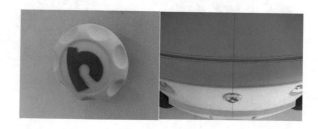

Fig. 14.8 Seat layout

14.6 Future Feature: Exchangeable Exterior

After deciding to use textile as RimOnO's body material, we wanted to make the exterior changeable–just like changing clothes. Although no changeable textile exterior was incorporated in the RimOnO Prototype 01, Kota made different images of RimOnO's exterior (Fig. 14.9).

With such an exchangeable exterior, users can choose from a variety of designs just like with smartphones, thus creating opportunities for third-party designers and vendors. Business users can create their own exterior design to promote their products/services while driving their RimOnO in the streets and return to the original exterior after business hours. Shared RimOnOs could have original local designs to differentiate cars from other cities.

Fig. 14.9 Exchangeable exterior

14.7   Launching the First Prototype

After being launched in May 2016, the RimOnO Prototype 01 received wide media coverage in Japan, including NHK, TV Tokyo, Nippon Television, TV Asahi, Nikkei, Forbes Japan, and many others.

We received many emails and calls from as far away as Thailand, Sweden, and the U.S. from potential users about when and how to buy it. Most of the inquiries in Japan came from two groups: (1) men over 75 who are contemplating renouncing conventional cars; and (2) women around the age of 50–60 who lack the confidence to drive conventional cars but need a vehicle to take their physically disabled parents to the hospital.

Unfortunately, our project has been suspended because we failed to get sufficient funding for mass production and because regulations that would have permitted permit tiny two-seat electric cars were never enacted by the government.

14.7.1   Changing City Streets

The twentieth century's motorization has introduced many benefits to society. However, it also created significant environmental problems, traffic accidents, and

Fig. 14.10   Streets in future

congestion. Most city streets are merely "driving lanes" for cars and pedestrians have been chased to the sidewalks.

If RimOnOs were driven in city streets, pedestrians wouldn't be relegated to the sidewalks by such tiny and slow vehicles. Streets might once again become fun places to walk around, populated by kawaii cars with interesting textile coverings on the body of each RimOnO. There would be more open cafés, restaurants, and shops because the streets welcome pedestrians.

As we enter the 2020s, we have to reimagine how to use our streets. Personally, I believe that most street spaces should be returned to the local citizens as places to communicate, where kids can play, people can go shopping, and community events can be held. Even if RimOnO fails to be mass-produced, it would still be wonderful to experience the changes that could happen in the streets in many cities to make them more pedestrian friendly (Fig. 14.10).

Part IV
Epilogue

Chapter 15
Editor's Summary

Michiko Ohkura

Abstract This chapter summarizes this book and the future of kawaii engineering.

Keyword Kawaii engineering

More than 10 years have passed since I started my research on kawaii engineering. I am editing this book to summarize our research results with other researchers.

Chapter 1 introduces the background of kawaii engineering.

Chapter 2 describes our systematic measurement and evaluation methods for kawaii industrial products in terms of such physical attributes as color, shape, material perception (visual texture and tactile texture), and sound. As I have already described in Chap. 1, actual industrial products consist of various combinations of these physical attributes, and such impression as kawaii should be determined from interactions with them. However, for engineering research, basic experimentation with each single attribute is the appropriate first step. Although the research results might not be surprising, as engineering research, systematic procedures are critical even though the target concept, kawaii, has some ambiguities.

We researched affective evaluations for the material perception of bead-coated resin surfaces using visual and tactile sensations. Chapter 3 introduces research that focuses on kawaii.

Chapter 4 describes kawaiiness in motion by K. Tomiyama and his group.

Chapter 5 introduces our various evaluation research using many different biological signals. Chapter 6 describes our kawaii evaluations using eye tracking. Recently, not only facial expressions and voices but also such biological signals as EEG and ECG are employed for affective computing. We have been employing these signals for affective evaluations for more than 10 years. In addition, we have successfully classified both exciting kawaii and relaxing kawaii using ECGs.

In this research, most participants were Japanese males and females in their twenties who are very familiar with kawaii because it is a positive value for Japanese in that age range. We employed virtual environments in most of our research to show kawaii objects to the participants. I received my Ph.D. using virtual environments

M. Ohkura (✉)
Shibaura Institute of Technology, Tokyo, Japan
e-mail: ohkura@sic.shibaura-it.ac.jp

© Springer Nature Singapore Pte Ltd. 2019
M. Ohkura (ed.), *Kawaii Engineering*, Springer Series on Cultural Computing, https://doi.org/10.1007/978-981-13-7964-2_15

in 1995 and employed them in much subsequent research because employing virtual objects is useful for a systematic engineering approach. However, because of the rapid progress of 3D printer technology, the systematic manufacturing of actual objects using 3D printers has been greatly simplified. Therefore, the combination of virtual appearance and actual tactile feeling in Chap. 3 is a good example of an appropriate combination of virtual and real environments.

Chapter 7 introduces scientific approach to kawaii called "Meaning of kawaii from a psychological perspective" by H. Nittono.

The chapters in Part I provide basic information about kawaii engineering for researchers with manufacturing and service providers.

Chapters 8–14 introduce various applications of kawaii engineering. The research described in Chap. 8 is a collaboration with a cutlery manufacturer. Chapter 9's proposal is a collaborative research with an optical equipment manufacturer. Chapter 10 introduces research on affective values for Saudi women using an extended ground theory approach. Chapters 11 and 12 introduce comparative research between Thai and Japanese that employed machine learning models to express kawaii feelings. Chapter 13 compares Chinese and Japanese attitudes toward it. Chapter 14 introduces a new last-mile mobility, a small car, manufactured by an entrepreneur. Kawaii is not an important value in Saudi Arabia, even for females. It is also not critical for Chinese males. On the other hand, kawaii is a critical affective value for both Thai males and females. For a Japanese male entrepreneur of a new small car, kawaii is most essential. Kawaii is a common affective value for young generations in many countries in East Asia. I am writing this chapter in Paris, where I stumbled across a kawaii stationary shop. France has embraced various Japanese affective values for over 200 years. And recently, the country has also begun to welcome kawaii products. As I already mentioned in Chap. 1, the Japanese word "kawaii" and its concept might be shared worldwide to encourage peace and sustainable development goals (SDGs) in the twenty-first century.

Kawaii engineering remains in its early adolescence. However, we achieved its modeling step using machine learning, as described in Chaps. 11 and 12. In addition, we proposed a new method to quantitatively clarify the contributions of physical attributes using deep CNN, which was described in Chap. 12.

The chapters in Part II provide examples of kawaii engineering for researchers and give hints for developing kawaii products for manufacturing and service providers.

I hope this book fuels an increase in the number of the researchers who address kawaii engineering to promote this research field. I hope it raises the numbers of manufacturers and service providers who will pay more attention to make their products more kawaii.

Index

Printed in the United States
By Bookmasters